Running Dry

PINPOINTS

Pinpoints is a series of concise books created to explore complex topics by explaining key theories, current scholarship, and important concepts in a brief, accessible style. Each Pinpoints book, in under 100 pages, enables readers to gain a working knowledge of essential topics quickly.

Written by leading Rutgers University faculty, the books showcase preeminent scholars from the humanities, social sciences, or sciences. Pinpoints books provide readers with access to world-class teaching and research faculty and offer a window to a broad range of subjects, for a wide circle of scholars, students, and nonspecialist general readers.

Rutgers University Press, through its groundbreaking Pinpoints series, brings affordable and quality educational opportunities to readers worldwide.

The series comprises the following five volumes:

Deborah Carr, *Worried Sick: How Stress Hurts Us and How to Bounce Back*

Nicole R. Fleetwood, *On Racial Icons: Blackness and the Public Imagination*

James W. Hughes and Joseph J. Seneca, *New Jersey's Postsuburban Economy*

Toby Craig Jones, *Running Dry: Essays on Water, Energy, and Environmental Crisis*

Charles Keeton, *A Ray of Light in a Sea of Dark Matter*

Running Dry

Essays on Energy, Water, and Environmental Crisis

Toby Craig Jones

Rutgers University Press

New Brunswick, New Jersey, and London

LIBRARY OF CONGRESS CONTROL NUMBER: 2015933181

ISBN 978-0-8135-6996-3 (pbk.)—
ISBN 978-0-8135-6994-9 (ebook (Web PDF))—
ISBN 978-0-8135-7545-2 (ebook (ePub))

Visit our website: http://rutgerspress.rutgers.edu

Manufactured in the United States of America

Contents

Introduction

OUR CONTINUED EMBRACE OF and dependence on cheap and easily accessible carbon-based energy—coal, oil, and natural gas—is no longer sustainable. In spite of the claims of a handful of politically or commercially motivated skeptics, the evidence is clear that the rapid consumption of carbon energy over the past 150 years has aroused the specter of environmental calamity. We live in a moment in which the range of threats is breathtaking. At the global level, the most pressing danger is the impact on the climate of the emissions from so much carbon consumption. A combination of gases from the firing of coal and oil, particularly in heavily industrialized and wealthy communities in the United States and Europe, has fundamentally altered the world's climate. Human behavior has always affected the environment, but, as the historian Dipesh Chakrabarty has elegantly argued, we have become geological actors on a scale previously the province of cosmic, solar, and volcanic forces. The results have been stupefying, ranging from the rise of superstorms and heavy weather in some places to drought in others. Average global temperatures are rising, although the atmospheric effects of warming have also produced record cold. Warmer patterns have accelerated the melting of the polar ice caps and rising sea levels.

The impact of carbon-based energy on the climate demands critical scrutiny, activism, and a commitment to finding ways of living that do not depend on fossil fuels. It is impossible to predict the outcomes that shifting climate patterns will produce. Those that are already evident are frightening enough. The possibility that rising sea levels could eventually destroy some of the world's coastal cities should stir all of us to do more. The past decade has seen meaningful mobilization against carbon-based energy. Environmental and political networks that have raised the issue of climate change and the broad environmental effects of our rapid industrialization have become mainstream. Public pressure has been significant enough to compel policymakers to do more, or at least to acknowledge that more needs to be done. And yet there is little evidence that global consumption of oil, coal, and gas is set to decline in any meaningful amount in the near future. Both production and consumption of oil and coal continue at alarming rates.

There are other dangers too, which are often ignored amid the urgency and anxiety that have accompanied revelations about climate change because their scale appears less ominous. They should not be. Indeed struggling to address and perhaps roll back climate change should not come at the expense of awareness of and concern about carbon energy's *other* environmental and political wages. It should also not come at the expense of examining why we have privileged the need for cheap carbon-based energy over more important natural resources, such as water. Set against our romance with oil, the kind of living that carbon energy makes possible, the past four decades have seen consumers as well as corporate and political interests prioritize access to carbon-based energy often at the expense of protecting the environment and political liberties.

Since the closing decades of the twentieth century, our collective desire to do whatever is necessary to maintain cheap and abundant oil has caused us to pollute our water, transform the climate, endanger environments, and allow a powerful energy industry to bend political systems to their will.

When it comes to the environment, since the 1970s and especially since the recent rush on energy that has taken hold in the United States, water has begun to disappear in some places, is being turned into a threat in others, and is being siphoned away or poisoned by modern industry, particularly the global energy industry. Access to safe, cheap water is becoming limited. The extent to which the world is running dry is the result of neglect or the belief that the tools of the market and business, what ostensibly drives the need for energy, can and should work for water as well. They do not, and the damage done by this belief is increasingly clear. In the American West drought has become a persistently stubborn problem; in parts of Africa and the Global South communities have lost access to water, which has been purchased by Western companies and Persian Gulf states. Carbon energy's impact on politics has been equally corrosive, poisoning democratic possibility and enriching an industry that has twisted science to serve its interests and undermined the power of citizens.

There are many reasons why energy has taken priority over the environment and political freedoms simultaneously. Oil and coal have been central to the rise of modern society, fueling the emergence of industry and transportation and making possible the comforts that we rely on. Carbon-based energy has brought unprecedented material comfort and health to the wealthy. But other factors that explain energy's privileged status are less clear. There has been no greater consumer of oil, coal, and gas than the United States, which is home to the world's most powerful corporate interests that have sought to extract these resources. Since the 1970s energy has become the focus of a particular national anxiety and commitment, one rooted in a sense of crisis and uncertainty that settled in the late twentieth century's desire to maintain reliance on carbon. The result has been that, collectively, Americans and their political leaders, encouraged by profiteers, have built a

Figure 1. Fracking across much of Colorado's Rocky Mountain Front Range takes place very close to homes. This drilling rig is in a neighborhood in Carbon Valley, Colorado. Photograph by Sandy Russell Jones.

political-economic order that puts energy at the center of its existence, growth, and security. Everything else, including the environmental effects of our modern energy system, has become secondary.

In the three essays that follow, I explore these themes broadly, making arguments for issues that I believe we need to struggle with more urgently. The essays address three main issues. In the first, I examine the scale of the U.S. energy industry and outline how the industry took precedence over preserving water and the environment in the 1970s. The reasons are rooted in anxieties that crystallized during the energy crisis of that decade and have endured ever since. The impact of this has been global. America's wars in the Middle East have been directly linked to the terms and ways of thinking about energy that were put in place in the 1970s.

In the second essay, I examine the impact of the energy industry itself, particularly the current wave of hydraulic fracturing, a technological process that has revolutionized how energy companies extract oil and gas, on the environment and public health in several parts of the United States, especially Colorado, home to one of the biggest fracking booms in the country. What is most alarming about the relationship between the current boom and the environment and public health is that clear scientific evidence of the risks and toxic effects of drilling for oil has been either ignored or undermined. The science has been politicized and the political regulatory system that should be managing the risks has instead been subordinated to the energy industry itself.

The last essay turns away from the water-energy nexus and examines the political consequences of relying on carbon energy, particularly the antidemocratic effects that energy has wrought historically. Here I argue that it is useful to consider the ways in which producing oil has undermined democracy globally, especially in the Middle East, and in the United States, where antidemocratic politics are thriving in energy-rich communities. More important, though, is the argument that democratic politics in the United States is hardly assured; it must be struggled over and protected against the convergence of narrow corporate and political interests. There is reason for optimism, as networks of activists who do prioritize their local water resources and local environments have stubbornly and, albeit in limited ways, successfully resisted the encroachment of Big Energy. Their example should give us hope that protecting water and the environment can be restored as political priorities.

There is much that is missing in *Running Dry*, including close scrutiny of class and race. (It is often the case that the least privileged are the most negatively affected by our collective energy and environmental choices.) These essays are conceptualized and pitched broadly. Where useful or evocative, details are offered, but it is necessary to look at these issues through a wide lens. I often refer to "Big Energy" or the "energy industry" in the pages that follow because, although some energy companies are worse than others, they are all engaged in a pursuit with no happy ending. We need to see them as collaborators whose work is at odds with sustainability and long-term environmental and public health.

Influenced by my past academic work on oil and water in Saudi Arabia and the significant time I have spent in energy-rich communities in Colorado, New Mexico, Montana, and Pennsylvania over the past few years, these essays are meant to be informed provocations and calls to urgent attention. The essays blend empirical work with storytelling, historical thinking with journalism. My hope is that my attempt to portray this complicated and disturbing situation will encourage readers to make big connections. Along with my struggle to make sense of things, the work here relies heavily on the observations and activism of a broad network of people who live in communities that are most directly affected by the privileging of energy over water, over the environment, and over the climate.

I stop short of making particular policy recommendations, which I believe requires thinking that is not grand or ambitious enough and that preserves, even in limited ways, the existing order. Ending an oil-dominant and carbon-dominant energy regime requires a radical recommitment, not arguing for gradual tweaks to a broken system that elevates profit over communities. We do not need a better oil-based energy regime that gradually shifts us in a new direction. We need a new system and a commitment to providing the resources necessary to pursue it. Skeptics will almost certainly dismiss the cost and feasibility of what they will call a naïve approach in which the consumer will bear too high a burden. However, the unfortunate and increasingly clear reality is that the cost of the status quo will be even higher.

The title *Running Dry* does not mean to suggest that the essays that follow are about actual scarcity, about either oil wells or water aquifers drying out. Still, it is about loss, about both environmental loss and the loss of political possibility. When I set out to write this book I was determined to comment broadly on the ways water has become endangered and that safe water is and will be increasingly hard to secure. As projects tend to do, this one changed. But I have held onto the title. Threats to safe water remain urgent, and my concern with this danger underlies much of what follows, but focusing only on what is happening to water does not fully capture the scale of what is being lost. *Running Dry* is much more about the history of energy and the power of carbon energy to shape outcomes that are environmentally unsustainable, but that are also politically and economically ominous. The loss implied in the title, then, is meant to cover a range of concerns, from environmental loss to the forms of politics that carbon energy creates as well as those it has made impossible. The title is also a metaphor meant to suggest that the current system of energy on which we depend is running us dry, corroding the very conditions of happiness, health, and security that we claim it is intended to serve in the first place.

Chapter I

Choosing Energy

In August 2014 New Jersey governor Chris Christie quietly vetoed bipartisan legislation that would have prohibited the state from accepting wastewater produced by the extraction of gas and oil in neighboring states, particularly Pennsylvania, home to one of the country's most intense energy booms. Christie did not explain his decision, except to note that refusing the wastewater and finding ways of disposing it was at odds with the U.S. Constitution's rules governing commerce. There was no substantive discussion on the governor's part about the underlying concern that the state legislature sought to address, which was that wastewater from the process called hydraulic fracturing, or "fracking," is laced with toxins like benzene as well as radioactive elements, which, should they be brought into the state and disposed of locally, constitute a clear threat to New Jersey's public and environmental health.

It was not the first time the state legislature had attempted to address fracking. In 2012 legislators passed a bill that sought to ban it completely. It was a largely symbolic effort, since New Jersey is not home to rich energy deposits like its neighbors. Even so, Christie convinced the legislature to pass only a one-year ban instead, pending further study of fracking's dangers.

There are powerful reasons for concern about the dangers of treating and disposing of fracking waste. New Jersey has a dreadful legacy of toxicity and polluted landscapes. Its most dubious distinction is being home to the most Environmental Protection Agency toxic Superfund sites in the United States, places that are so contaminated that they are designated as particularly dangerous and marked for cleanup. Avoiding adding to the state's toxic woes should be a clear priority. Beyond the history of pollution and toxic dumping and the consequences of the state's industrial rise and fall, there are urgent concerns that are particular to fracking in Pennsylvania and the wastewater it produces.

Part of the concern is volume. Home of the Marcellus Shale Formation, a geological formation that contains trillions of cubic feet of natural gas, Pennsylvania is drilling and fracking for gas and oil on an unprecedented scale. Fracking involves injecting millions of gallons of water at high pressure into rock or shale that contains trapped gas and oil. The water has been laced with sand and chemicals to increase its destructive power and prevent blockages in the wells. The injected water helps break apart shale rock and frees gas to travel back up the well. Around 80 percent of the water injected into wells travels back to the wellhead, where it is temporarily stored on site and then trucked away for disposal. Because it contains chemical and other elements, water used for fracking is unsafe simply to dump in local rivers. It must be treated in order to remove as many of the toxins as possible before its final disposal.

The recent energy boom, which has seen fracking for oil and gas rise dramatically in the past decade in Pennsylvania and other parts of the United States, has produced much more wastewater than ever before. Since 2004 the volume of Pennsylvania's wastewater has increased by perhaps as much as 570 percent.[1] Because the energy industry does not disclose details about the quantity of water it produces and uses, this figure is an estimate. But it is conservative to suggest that tens of millions of gallons of wastewater are now produced annually. Drillers use as much as four million gallons of water each time they frack a well. With thousands of gas wells in operation in Pennsylvania, and many more being planned, the volume is significant. Indeed it has overwhelmed Pennsylvania's capacity to manage it. Brian Lutz, a scientist who studies fracking and water, remarked in an interview in 2013 that channeling wastewater through treatment facilities had "been Pennsylvania's go-to method for decades" but that "these systems [are now] being overwhelmed. They were just taking too much waste, leading to water quality problems," and "there simply isn't [enough] disposal infrastructure in place."[2]

Pennsylvania's struggle to manage this massive volume has led drillers to seek more distant sites of disposal. Over the past decade, as wastewater levels have risen, Pennsylvania-based drillers have shipped or tried to ship their waste to New York, West Virginia, Ohio, and even Michigan. Aside from volume and the problem of infrastructure, there are more pernicious and dangerous reasons why Pennsylvania is struggling to manage the wastewater problem. Most important are the environmental and health dangers posed by the water itself and the invisible threats that inhabit it.

The energy industry has mostly resisted disclosing the makeup of the cocktail of water, chemicals, and sand that it blasts into its fracking wells. As a result of recent pressure and scrutiny, driven by anxieties about their health from those who live near fracking activity, some companies have made available limited information about their chemical use. Fracking water, also known as produced water, contains a range of hazardous and carcinogenic materials, including benzene, arsenic, and various acids. In spite of protests from the energy industry that its practices are safe, there is increasing evidence of pollution from spills, from the seepage of fracking water underground, and from industrial negligence. (These patterns and the politics around them are examined in more detail in the second essay of this volume.) The energy industry's habits in Pennsylvania and its reliance on toxic water to maximize the extraction and production of gas are not exceptional. What is pumped into Pennsylvania's wells resembles similarly produced water elsewhere.

What distinguishes Pennsylvania, and thus the character of the threat, from most other centers of oil and gas extraction is the amount of radium and other radioactive elements that flow back out of the well with the wastewater. Radium is naturally occurring, a misleading point often offered up by energy companies that seek to downplay the risk. Although radium is indeed naturally occurring, it would normally remain hidden deep underground if not for fracking.

Fracking's radioactive consequences have been reported across the United States, including in West Virginia, North Dakota, and Colorado. But drilling in the Marcellus Shale is particularly likely to create radioactive dangers. In 2011 the U.S. Geological Survey published a report arguing that the levels of radioactive radium, uranium, and thorium in wastewater from the Marcellus Formation were far higher than elsewhere. This higher concentration likely comes from large saline water aquifers in the region's Appalachian Basin.[3] There are several dangers from high levels of radioactive materials dredged up from fracking. Water treatment facilities can theoretically remove them, although observers believe doing so is made more difficult as the volume of water being treated grows exponentially. Volume, then, matters. In addition, while humans may not come into direct contact with or drink radioactive water, there are still significant environmental risks that threaten public health. Like other toxic materials, once produced water settles into the ground or leaks into freshwater and is ingested by fish, livestock, or other animals or is exposed to plants, the radiation it emits can alter the plants' and animals' biological makeup. Over time the cumulative effect can be quite dangerous.

There are direct threats to people who come into contact with radium. Susan Phillips, a journalist based in Pennsylvania who has written extensively about the risks of fracking, notes that while the Environmental Protection Agency (EPA) concedes that human bodies can "eliminate the bulk of radium" that gets ingested or inhaled, any exposure nevertheless raises the likelihood of lymphoma, bone cancer, and leukemia and other blood-related diseases. She quotes the EPA, which states that "these effects take years to develop. External exposure to radium's gamma radiation increases the risk of cancer to varying degrees in all tissues and organs."[4]

Pennsylvania's Department of Environmental Protection was sufficiently concerned about the increase of scientific evidence for the radioactive dangers of wastewater in 2013 that it commissioned its own study from the Atlanta-based waste management firm Permafix. Permafix issued a report in January 2015 that argued that the risks of direct radioactive exposure to the public or to workers at treatment facilities were limited. Handled correctly, Permafix claimed, radioactive wastewater and solid waste, like the mud or sludge that such water is often mixed with, have a low likelihood of significant danger. The report's conclusions hinged on safe handling. Permafix noted that spilling or negligent handling, which might dump wastewater into the environment inadvertently, would be cause for alarm.[5] The message is that while the waste itself is dangerous, proper management and expertise will assure limited risk.

There are several reasons to be skeptical of Permafix's conclusion. The company's faith in technological management and the power of proper handling deserve scrutiny. Faith in technology and the power of experts to handle dangers has a long history in the United States and elsewhere, despite how frequently they fail. The energy industry is no exception to failure, including in managing the waste it produces. The reality is that as drilling and fracking have intensified in scale, so too have spills and leaks and accidents. In 2014 there was on average at least one reported wastewater spill a week in Pennsylvania.[6] Spills are supposed to be reported directly to state authorities, a regulatory demand that some energy companies observe. Others do not. The scientific and activist collective at Fracktracker.org uses community resources and crowdsourced reporting to build empirical data and mapping analysis of wastewater contamination and other energy industry practices that threaten the environment and public health. The evidence is overwhelming that, even if well intended and committed to caution, the energy industry is confronted with so much waste that it is impossible to handle it all safely and effectively. In early January 2015 almost three million gallons of wastewater produced by the oil and gas industry in North Dakota spilled through a broken pipeline into a creek system just fifteen miles outside the city of Williston. The size of the spill is remarkable, although it is not unprecedented. Millions of gallons of waste have spilled into some of California's freshwater aquifers in the past few years. Perhaps what is most remarkable about the spill in North Dakota is that it took Summit Midstream Partners, the company that operated the leaking pipeline, several weeks to determine the size and scale of the spill.[7] When it comes to the scale of the threat and the magnitude of the potential dangers, even the energy industry is not always fully aware.

There is also evidence that the regulatory and safety burden state and national environmental agencies impose on energy companies is often disregarded. Not all energy companies seek to outmaneuver expensive environmental and health regulations, but some do. Illegal dumping and attempts to escape regulatory oversight have been reported all over the country. Scott Radig, who oversees North Dakota's waste management operations at the state's Health Department, told a reporter at Bloomberg, "Some [waste] ends up in roadside

Oil and Natural Gas Exploration

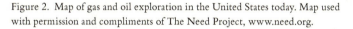

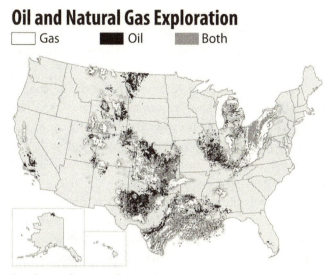

Data: Energy Information Administration

Figure 2. Map of gas and oil exploration in the United States today. Map used with permission and compliments of The Need Project, www.need.org.

It has long been the case that those who aspire to political office, particularly the presidency, cozy up to the moneyed classes. And there are fewer classes more wealthy and powerful than the forces behind Big Energy. Christie has left little doubt that he is committed to supporting the energy industry's interests, even at the expense of public and environmental health.

In December 2014, just months after vetoing the wastewater ban, Christie went to Calgary, Canada, where he met with and expressed his devotion for the chief executive of the corporation that seeks to build the Keystone XL pipeline from Canada to the Gulf of Mexico. Christie delivered remarks, almost certainly as a booster for the industry and the XL pipeline, at the Calgary Petroleum Club. It is important to consider why Christie, known as a brazen political operator, chose this venue and why energy receives the kind of support that was on display in Calgary. Doing so will also allow for some reflection on why energy and supporting the pursuit of it has been privileged over concerns about public health, the environment, and living with the potential risks of our dependence on carbon-based fossil fuels.

The prospect of the Keystone XL pipeline has generated both massive opposition and massive support. Advocates argue that the pipeline is essential for making up to 830,000 barrels of Canadian oil available for consumption in a world that has a seemingly unlimited thirst for petroleum.[10] Its most powerful and outspoken backers are fellow members of Christie's Republican Party, which just a month earlier enjoyed a sweeping electoral victory, gaining seats and power in the U.S. Congress. It was widely anticipated that among the new majority's first acts once seated in January would be to put forward congressional support for building the pipeline. The pipeline's opponents have doggedly cited the dangers it poses to the environment, particularly from spills and leaks, and have questioned its ability to satisfy energy demand or create jobs.

These disagreements over the pipeline adhere to familiar partisan political fault lines between Republicans and Democrats, especially after President Barack Obama stated in January 2015 that he would veto any legislation supporting its construction. He did so in February. The reality is more complicated. Obama has not opposed the pipeline because of its environmental risks, at least not publicly. In fact he and other major Democrats are as enthusiastically supportive of the energy industry in general as Christie is. His opposition to the legislation has more to do with Beltway politics than environmental principle. For now, however, Obama is waiting for the U.S. State Department, whose involvement is mandated in matters like the Keystone XL pipeline, which crosses national borders, to make its determination before he decides his level of support.

Christie has left little doubt about his support. In Canada he remarked, "On the merits, Keystone should have been approved a long time ago. . . . It is time—well over time—to get this done." He dismissed concerns about safety, saying, "You know, in the United States, we already have over 2.2 million miles of pipeline. Canada has tens of thousands of miles of pipelines. In both cases, the safety record is sound."[11] His enthusiasm was called into question a month later, when 40,000 gallons of oil spilled into the Yellowstone River in Montana, rendering water undrinkable for thousands of residents near the town of Glendive.[12]

It is tempting to criticize Christie for traveling thousands of miles from New Jersey to stump for the energy industry, to argue that he is grandstanding on an issue that is increasingly central to national Republican Party operatives and yet distant from the concerns of those he actually represents, or to dismiss his behavior as opportunistic. After all, he has a demonstrated record of carefully crafting his political image and recklessly pursuing self-interest. But such criticism would be mistaken. Equally mistaken would be to view Christie's energy politics through the lens of contemporary political partisanship in the United States. While the Republican Party has been particularly enthusiastic about the Keystone XL pipeline, the reality is that support for and from Big Energy transcends party affiliation. Democrats are perhaps not as callous or outspoken in their public support for the domestic energy industry, largely because many Democratic officials recognize that part of their base opposes the industry; nevertheless they support it fundamentally. The 2014 Democratic-controlled Senate fell only one vote short of authorizing the construction of the pipeline.

With all of its wealth and power to shape campaign treasure chests, the energy industry has likely purchased much of its support, although its success has not been due solely to corruption and the influence of money. The reasons for the industry's success are rooted in the particular ways that energy and especially oil were prioritized and privileged in the United States in the late twentieth century.

I suggest that understanding why supporting energy seems to consistently trump protecting the environment requires a look at developments in the late twentieth century, particularly the moment when the possibility of setting American energy policy on an environmentally friendly course was lost. Since the 1970s American policymakers and the public have struggled to reconcile contradictory interests: the country's dependence on oil and a growing concern for the environment and public health. Presented with the choice of protecting the environment and pursuing potentially more expensive but unquestionably healthier and more sustainable energy choices, or continuing to be dependent on oil, gas, and coal, Americans have mostly chosen the latter. Protecting the environment has been rendered a secondary concern at best. Particularly with the rise of an antiscience political class that has sought to undermine efforts to stem climate change, environmental dangers and concerns have been dismissed as unfounded. This latter development is often a corporate-backed assault, but there are other forces that explain why carbon-based energy has retained primacy at the expense of other possibilities and at the expense of the environment. Some of these have to do with the social and cultural consequences of oil in the early twentieth century and their lasting legacy on the ways we live in the world. Others reflect a particular kind of national politics and anxieties about security that took shape in the closing decades of the century.

OIL, OF COURSE, has been central to the making of modern America since the closing decades of the nineteenth century, when it emerged as a critical source of power for industry and transportation. It has been the quintessential industrial commodity ever since. One of oil's advantages in the United States was that for the first two-thirds of the twentieth century, it was plentiful at home. The oil patches of Pennsylvania, Texas, Oklahoma, Louisiana, and California produced enough to satisfy the needs of rapid industrialization and the resulting social and technological changes. Lighter and more efficient than coal or wood, it quickly displaced potential energy alternatives. For oil producers, both large and small operations that sought to profit from drilling and marketing American oil, the problem was never that oil was scarce. Rather, much of oil's early history in the United States and even globally was marked by concern that there was too much of it. So much easy oil facilitated rapid dependence. Just as important, it also rendered oil an afterthought, a source of power readily available and with so little effort that its abundance and cheap cost were taken for granted.

But by the end of the 1960s, America's appetite for oil ran up against lagging domestic production. The golden era of American oil was over. It is not that America had run out of oil but that the easy-to-get oil was being depleted. Accessing deeper oil, trapped in shale formations and located in remote outposts, was prohibitively expensive, especially considering that there was plenty of oil available globally. The decline in American resources was unsettling for policymakers, who worried about the potential leverage that foreign suppliers might command over the U.S. economy.

Of course, not everyone shared this anxiety. As late as the late 1960s it was still the case that the largest supplies of global oil were under Western corporate dominance. Powerful European and U.S. oil companies had expanded globally in the early twentieth century, even helping to shape the borders of the Middle East, where the world's largest oil supplies exist, in the interest of securing a foothold there. U.S. and European corporate dominance would pass by the

1970s, however, as Arab states began to take direct control over their own oil, underscoring a deepening sense of anxiety that was beginning to become clear in the 1960s. Much of the oil consumed in the United States was still domestically produced, and most of the important energy companies, at least those that refined, transported, and made oil and gas available for consumption in the United States, were American. But not all American demand could be met through American sources. Gaps in supply had to to be filled from abroad, leading to concerns about potential shortfalls and anxieties that oil would be increasingly scarce and hard to come by. Alarm about scarcity and having to rely on foreign production, on states in the Middle East to extract and make available the crude oil that would be turned into gas, to meet domestic needs was growing.

No longer able to rely fully on its own supplies, the United States was settling into an era that Presidents Richard Nixon, Gerald Ford, and Jimmy Carter, along with a generation of policymakers and consumers, characterized as one of "crisis." The long energy crisis of the late twentieth century was first given expression by Nixon, who began to be alarmed by rising energy prices and the specter of scarcity in 1968. Three years later he outlined what became a central occupation of his administration and those of his immediate successors, arguing that "a major cause of our recent energy problems has been the sharp increase in demand that began about 1967. For decades, energy consumption had generally grown at a slower rate than the national output of goods and services. But in the last four years it has been growing at a faster pace and forecasts of energy demand a decade from now have been undergoing significant upward revisions."[13] Nixon called for a sweeping new approach to energy, a way of thinking about oil, its production, and consumption that was increasingly complex and tinged with urgency and fear. He also called for the energy industry, and its protection, to be better integrated into the American political system, which required greater federal oversight.

The long energy crisis, as the Yale University historian Paul Sabin has written, developed at the same time as a growing set of concerns about environmental protection. But while environmental protection would grow teeth in the 1970s, this development did little to alter American dependence on petroleum. Some growing environmental concerns were connected directly to oil, as was the case after the 1969 oil spill off the coast of Santa Barbara, California, which, along with hundreds of smaller spills by the early 1970s, brought attention to the risks of extracting oil and transporting it by pipelines. More generally, though, emerging environmental politics were based in apprehensions about industrial and chemical dangers lurking in landscapes and bodies. Rachel Carson's book *Silent Spring*, published in 1962, about the dangers of pesticides to ecological systems and public health, kindled concern about unregulated industrial agriculture. Carson's call for greater efforts to protect the environment received a boost with the 1969 Santa Barbara spill and when the Cuyahoga River near Cleveland, "awash in refinery waste and other debris," caught fire six months later.[14] By 1970, when millions of people marched in support of the first Earth Day, it seemed that the environment's moment was at hand.

National attention and concern produced what appeared to be meaningful policy outcomes. President Nixon created the Environmental Protection Agency in 1970 and ushered a series of water protection measures through Congress early in the decade. Confronted with declining oil production in the early 1970s and growing national concerns about dependence on foreign oil, Nixon sought to reconcile the energy crisis with the new environmental politics. In April 1973 he asked for a more robust and careful approach to energy, including increasing domestic energy production wherever possible (opening

Alaska to exploration), energy conservation (lowering consumption), and embracing the need to import more oil. This
last point was particularly tricky, as domestic producers had
enjoyed protections from foreign competition, including caps
and taxes on imports. Nixon forged ahead, lifting tariffs on oil
on April 18, promising, "This action will help hold down the
cost of energy to the American consumer."[15]

In pushing for expanded access to foreign oil and for finding
more sources of domestic production, Nixon sought to straddle
the line between making more energy available and protecting
the environment. He was also "striv[ing] to meet our energy
needs at the lowest cost consistent with the protection of both
our national security and our natural environment." He suggested
the country's energy and environmental needs could be managed
together: "In determining how we should expand and develop
these resources, along with others such as nuclear power, we must
take into account not only our economic goals, but also our environmental goals and our national security goals. Each of these
areas is profoundly affected by our decisions concerning energy.
If we are to maintain the vigor of our economy, the health of
our environment, and the security of our energy resources, it is
essential that we strike the right balance among these priorities."[16]

Nixon's hope of managing an energy policy while protecting the environment ultimately failed. Over the next few
years efforts to regulate environmental protection would continue, but they did little to alter the central importance of oil.
Indeed oil and the demand to protect "access" to it globally
would supplant environmental concerns. The terms in which
this divide took shape involved the stark language of crisis and
the rise of a new kind of political emphasis on energy security
and the pursuit of energy independence, a way of thinking
about energy policy that aspired to an era of oil plenty lost with
declining production in the late 1960s. The country's leaders
talked of an energy crisis, which they explained as a looming

danger that sufficient oil would not be available to meet American needs from secure sources.

It is worth reflecting on this. The anxieties around energy in the 1970s had little to do with material, environmental, or political consequences of being dependent on oil itself. Rather Americans feared being dependent on oil from foreign sources. The emerging sense of crisis could have generated a meaningful push toward alternative sources of energy. President Nixon, those around him, and observers everywhere began to talk openly about pursuing non-carbon-based energy sources, but the pursuit was almost entirely rhetorical. Very little was done to accomplish a break from oil.

It is not hard to understand why. Oil had become so dominant, had shaped social and other relations in such fundamentally meaningful ways, and even at the beginning of the crisis was still cheap enough that actually pushing an alternative energy agenda would have come at considerable cost. The early 1970s clearly marked a moment when an opportunity was lost, but it bears acknowledging how difficult ushering in a post-oil moment would have been. Consumers would have borne most of the expense of any transition. Just as important, the corporate forces behind oil and gas would have stubbornly resisted. And they had the power to mount significant opposition.

The oil companies were undergoing their own transition and expansion, steadily transforming into larger and more complex corporate entities, becoming Big Energy in the late 1960s with power not only over oil but increasingly over more disparate sectors of the energy industry too. Joe Stork, who wrote about the energy crisis in the middle of the 1970s, documented that besides owning large parts of oil production in the United States and globally, the energy industry moved into natural gas, coal, and nuclear power. With increasing control over a vast range of resources and markets, the energy industry, which sought profit over security, was a powerful barrier to any kind of post-petroleum transition. It remains so today.[17]

While many both inside and outside of the establishment recognized that the environment was imperiled, there was no widespread corresponding urgency about what it meant for the environment to be threatened, at least none that rose to the level of the crisis talk about energy. Because environmental damage was often hard to measure, sometimes taking years to manifest, and was not always clearly linked to social and economic needs, it was not viewed in terms as stark as oil and energy needs. Sabin has argued that activists and those most engaged in environmental matters made no particular effort to distinguish oil as exceptionally or particularly dangerous to the environment or to public health. It was enough to try to sustain a broadly conceived approach to the environment in which oil was one concern among many. Because of petroleum's pernicious impact on the environment ever since, and especially its climate effects, in hindsight this seems to have been a bad strategy. However, once access to oil rose to the level of national vulnerability and was marked as part of an unfolding crisis, there was little rhetorical or political space to single it out critically.

Crisis talk spiked in frequency and urgency in the early 1970s. Commentators speculated that there was potential for yet more serious trouble, especially as perceptions crystallized that the United States was increasingly dependent on and vulnerable to foreign supply shocks and that, with tensions growing in the oil-rich Middle East in particular, oil producers around the world that were critical of U.S. foreign policy might decide to use oil as a weapon. These worries seemed to be realized in the fall of 1973, when the most powerful oil producers in the Middle East announced an embargo against the United States. The combination of war and geopolitical anxiety that shaped the Middle East in 1973 and 1974 also left a lasting imprint on U.S. energy policy and the ways American policymakers and consumers would come to think about oil, and it helps explain why energy and oil would come to enjoy greater status than protecting the environment.

The oil embargo followed the outbreak of war between Israel and Egypt in October 1973. Hoping to regain Israeli-controlled territory in the Sinai Peninsula and pressure the United States into becoming an active broker in long-standing tensions with Israel over its regional role and the fate of Palestine, Egypt launched a surprise invasion against Israeli forces east of the Suez Canal. The attack, which caught Israel off-guard, led to initial battlefield success for the Egyptians and Israel's request for material U.S. support. The United States obliged, providing several billion dollars' worth of equipment in the midst of the fighting and prompting outrage from the region's oil producers. Led by Saudi Arabia, the Arab members of the Organization of the Petroleum Exporting Countries (OPEC) imposed an oil embargo against the United States and a handful of other Western countries. It lasted until March 1974.

The impact of the October War and the oil crisis of 1973–74 was wrenching and long-lasting. The embargo came at a difficult economic moment for Americans, who were struggling with inflationary pressures and generally anxious about economic malaise and energy and its availability. The embargo angered many Americans, who saw it as evidence of vulnerability. These anxieties were compounded in the fall and winter of 1973 and 1974, as consumers were confronted with long lines at gas stations and scarcity in the heating oil market. Fear of foreign oil power alongside frustration at not being able to access what was once so available helped shape what would become a deep antipathy toward Arab oil producers. It did not help that foreign producers were also raising their prices, led by Iran. Price levels, which had historically been controlled by oil companies and kept low, rose from $3 a barrel to over $12 in the spring of 1974. Prices would move even higher before flattening out a decade later.

However, the material impacts of the embargo have often been overstated. As Joe Stork, Timothy Mitchell, and others have demonstrated, the embargo was largely ineffective in keeping oil from the United States.[18] There was no actual shortage; the long gas lines were the result of Nixon's putting in place a rationing policy that limited sales and an overstressed refining capacity. These details have yet to be fully appreciated by historians. The dominant narratives that took hold during and immediately after the embargo placed blame on and directed anger toward the large oil producers in the Persian Gulf. Belief that the United States was a victim of avaricious oil sheikhs who aimed to expose and capitalize on American energy vulnerabilities has persisted ever since. Much is lost in this way of thinking, including the initial impetus for the embargo and rising prices: America's Middle East policy, its disregard for Palestine, and its political and material support for a rapidly militarizing Iran, which sought expensive American weapons and needed high oil prices to buy them.

In addition to growing anti-Arab sentiment, other responses to the oil embargo have shaped energy policy and the collective American embrace of oil, particularly at the expense of the environment. The first was Americans' belief in domestic oil scarcity and the corresponding vulnerability from having to rely on foreign oil. Nixon's response in the fall of 1973 was to accelerate the energy policies he had outlined the previous spring. To survive the crisis he imposed significant limits on consumption: "In order to minimize disruptions in our economy, I asked on November 7 that all Americans adopt certain energy conservation measures to help meet the challenge of reduced energy supplies. These steps include reductions in home heating, reductions in driving speeds, elimination of unnecessary lighting. And the American people, all of you,

you have responded to this challenge with that spirit of sacrifice which has made this such a great nation."[19]

Nixon also called for the country to become "energy independent," a vision that has remained central to how we think about energy in America today. Despite insisting just a few months earlier that energy policy be pursued with regard to environmental protection, Nixon now made the environment a secondary concern. Appealing to the symbolic power and cultural weight that Americans attach to the notion of self-sufficiency, he argued that the country's "overall objective" should be independence:

> From its beginning 200 years ago, throughout its history, America has made great sacrifices of blood and also of treasure to achieve and maintain its independence. In the last third of this century, our independence will depend on maintaining and achieving self-sufficiency in energy. . . . As far as energy is concerned, this means we will hold our fate and our future in our hands alone. As we look to the future, we can do so, confident that the energy crisis will be resolved not only for our time but for all time. We will once again have plentiful supplies of energy which helped to build the greatest industrial nation and one of the highest standards of living in the world. The capacity for self-sufficiency in energy is a great goal. It is also an essential goal, and we are going to achieve it.[20]

He proceeded to make the remarkably ambitious claim that with determined effort and careful planning, something he called Project Independence 1980, by the end of the 1970s "Americans will not have to rely on any source of energy beyond our own." The president was overreaching. But in the atmosphere of anger and fear, his ambitious gambit was well received.

The urgency of the moment and the scale of the oil crisis refocused the White House's and national priorities around energy and ensuring access to it. In the years and decades that followed the oil crisis, the pursuit of energy and the terms in which it was characterized singled it out as particularly central to the country's economic health and national security. Protecting the environment, though still urged by public officials and activists, never rose to the same level of interest.

Nixon's pursuit of energy independence became a central theme in U.S. politics in the late twentieth and early twenty-first century, crossing partisan political lines. After the Republican Nixon, President Jimmy Carter, a Democrat, asked Americans to practice conservation at home by turning down their thermostats. Although oil had been a national security issue at least since World War II, fears of U.S. economic vulnerability had intensified. Only by exploring for energy at home could Americans be safe.

There is a contradiction at the heart of this vision for energy independence, for U.S. policymakers ended up developing even closer ties to and strategic relationships with Iran, Saudi Arabia, and other energy-rich countries in the Persian Gulf. Over the rest of the 1970s the United States would sell billions of dollars' worth of weapons to the shah of Iran and the Saudi royal family. With the fall of the shah in 1979 and the Soviet invasion of Afghanistan, the United States accelerated its military support for allies in the region, especially Saudi Arabia, and became involved in a long war in the Persian Gulf.[21] The origins of this military commitment were outlined in the winter of 1980, when President Carter promised to use whatever means necessary to protect "vital resources" in the Persian Gulf. Carter's euphemistic reference to oil and American concerns about protecting access to it belie the underlying claims about energy independence.

Indeed the deepening ties to the Arab oil producers made clear a basic flaw in the pursuit of energy self-sufficiency: it is impossible, at least if the primary source of energy remains oil and other carbon-based resources. U.S. rates of oil and gas consumption outstrip that of every other society on the planet. In 2013 Americans used almost nineteen million barrels of petroleum products daily. In 2014 U.S. domestic production of oil totaled only about nine million barrels, with natural gas production adding a few million more, still well short of meeting basic demand. This gap was clear as early as the 1970s, and yet officials and influential policymakers have routinely referred to the need for energy independence ever since. Much of this has been political grandstanding, a way to tap into some mythological American triumphalism and resolve and to attract support at the ballot box. This was at the heart of Nixon's Project Independence 1980, perhaps much more so than any actual plan. During the 2008 presidential campaign, the Republican nominee, Senator John McCain, used "Drill, baby, drill!" as a rallying cry that became a mantra for the Republican Party.[22]

It is in this frame of energy independence that Chris Christie expressed his support for the Keystone XL pipeline in late 2014, against the backdrop of his veto of measures that would protect New Jersey's environment from the energy industry's waste. Following Christie to Canada, the *New York Times'* Michael Barbaro reflected, "As Mr. Christie weighs a presidential run, his trip here seemed calibrated to appeal to two crucial Republican constituencies: the elite corporate donors who loathe President Obama's inaction on the pipeline, and the grass-roots Republican activists who are convinced that it is vital to American energy independence."[23]

It is clear that energy independence has taken a commanding and uncritical hold over how many Americans think about oil and its importance to the economy and national security. The origins of this perspective are the uncertainty and anxiety that marked the mid-1970s. What exactly it entails is mostly mystified, especially the sheer scale of oil and petroleum products that energy self-sufficiency would require. In some ways it is the visceral power of the idea of independence more broadly, especially the ways it connects to notions of American strength and power, that is more important than the actual stakes involved in making sure energy resources are readily available. After all, in spite of rapidly growing rates of energy consumption in the United States, oil and gas have almost always been easily accessible and available. Even in the middle of the 1970s oil crisis, there was no real shortage of oil. It is as though the idea of scarcity and the possibility that American consumers might be cut off from oil are manufactured for purposes other than national security.

When it comes to the environmental dangers of so much dependence on oil and gas, the crisis of the 1970s and the politics of energy independence have had a pernicious effect. Prior to the oil crisis it seemed that environmental and energy policy, including the development of non-oil alternatives, would be developed together. The fallout from the oil crisis undermined this possibility.

THE 1970S' ENERGY AND ENVIRONMENTAL politics and the pursuit of energy independence have had other, subtle effects that remain in place today. Even those whose thinking has otherwise been progressive on the environment have struggled to overcome the power of old thinking about energy. In 2009 President Obama and the Democratic Party staked their energy policies to energy independence, much like their political rivals and predecessors. Shortly after being sworn in as president, Obama addressed what was then a pressing economic crisis and outlined an energy policy meant to steer the United States clear from future vulnerability. Obama stated, "At a time of such great challenge for America, no single issue is as fundamental to our future as energy. America's dependence on oil is one of the most serious threats that our nation has faced. It bankrolls dictators, pays for nuclear proliferation, and funds both sides of our struggle against terrorism. It puts the American people at the mercy of shifting gas prices, stifles innovation, and sets back our ability to compete."[24] He offered an energy policy that attended, at least notionally, to worries about too much consumption of oil, safeguarding against climate change, and establishing pathways to alternatives. He devoted a great deal of attention to curbing emissions by mandating stricter mileage requirements for automobiles. Yet even with this more complex approach to energy, with a view toward conservation and supporting green industry, his administration has also consistently backed fracking and, toward the end of his presidency, more drilling in places like the Gulf of Mexico. This is hardly the kind of break in energy policy that will steer the largest oil-consuming nation in a meaningfully new direction.

There is, of course, a need for a robust energy policy in the United States at the national level, although what has passed for talking about energy has almost always meant talking about oil and continuing dependence on petroleum. Obama's energy policy, while mostly well-intended and perhaps reasonable given the scale of American dependence on oil, was still little more than a better *oil* policy. As long as oil remains the dominant source of energy in the United States and in most industrialized countries, the environment, environmental protection, and related concerns about public health will be subordinate. This does not mean that they cannot be addressed or protected, only that the odds are long and that officials less inclined to listen to such concerns, like Chris Christie, will prove powerful obstacles to change. It might have served Obama better if he had thrown out the idea of energy independence and made a claim instead for rethinking what energy should mean today.

CHAPTER 2

Dangerous Water

THREE DAYS INTO AN unrelenting rainfall it was clear that communities along the Front Range of Colorado's Rocky Mountains were in trouble. On Monday, September 9, 2013, two powerful weather systems, one warm and wet, tracking up from the equator, and the other cold and dry and bearing down from the north, collided high above the mountains near the city of Boulder. Locked in an atmospheric embrace, the systems stalled over one of the country's most dangerous flood zones, dumping more than fifteen inches of rain over five days. The most intense rain fell from Wednesday into Thursday, dropping over nine inches onto already saturated ground.

The storm runoff in the mountains, especially atop Boulder Canyon, two thousand feet above and west of the University of Colorado campus, and Big Thompson Canyon, not far from the majestic Rocky Mountain National Park, turned the Front Range's creeks and small rivers into surging monsters. By midweek roiling floodwater poured down mountainsides, frothing along with bone-breaking and property-smashing debris: trees, collapsed bridges, remnants of destroyed homes, and slabs of asphalt wrenched from what were once winding roads. Colorado's governor John Hickenlooper declared virtually the entire Front Range, all or parts of fourteen counties, a disaster area.[1]

Heavy rain continued into the following weekend, stalling initial efforts at emergency management. Once the clouds cleared, the scale of the storm, the flood, and at least some of the human and physical costs became clear. Fast-moving water contributed to the deaths of at least eight people and caused up to $2 billion in damage.

Boulder County was among the heaviest hit. It was not the first time. The Rocky Mountain Front Range and Boulder have a history of catastrophic flooding. One of the earliest recorded floods in Boulder was in 1894, when a flooding Boulder Creek severely damaged the mining town. Another occurred in 1965. A decade later and farther north, heavy rain and flash flooding devastated areas along Big Thompson River and Canyon. In 1976 the Big Thompson flood killed 144 people.

The 2013 floods could have been much worse. Initial accounts figured the flood must have been a hundred-year event, which U.S. Geological Survey officials estimate have a 1 percent chance of happening annually. Calculating flood scales and probabilities is difficult. Based on at least one way of measuring, the 2013 flood was more likely a twenty-five-year event with a 4 percent chance of happening—still rare, but hardly unimaginable. Paul Danish at the *Boulder Weekly* observed that this determination reflected the volume of water surging through Boulder Creek at the height of the flood. Water in Boulder Creek normally flows at around 150 to 200 cubic feet per second. At its maximum in 2013, it measured around 5,300 cfs. To be a hundred-year flood, Danish wrote, requires "a flow of roughly 11,000–13,000 cubic feet per second."[2] These flow rates were not unprecedented. During the 1894 flood Boulder Creek had an estimated peak flow of 13,000 cfs, twice the rate that did so much damage over a century later. That a greater immediate disaster was averted had much to do with chance as well as broader climatic and environmental patterns.

The comparatively low rate of flow was remarkable considering the amount of rain that fell. Indeed it was the storm and not the flood that was historic. At first glance this is a paradox: How was it that so much rain did not produce a larger flood? The storm's total rainfall was the greatest ever recorded in the state's history, perhaps a thousand-year event. In less than a week rainfall in Boulder County exceeded its annual average. A smaller amount fell in July 1976—although twelve inches fell in twenty-four hours—that led to the Big Thompson flood and killed so many people. It mattered that in 2013 the rainfall was spread over a few more days. But it also mattered that the summer preceding the 2013 flood was characterized by warm, dry weather and a period of drought in much of the West. Diminished groundwater as well as low rates of flow higher in the mountains almost certainly prevented even faster-moving and more dangerous floodwaters.

Whatever the relative scale of the flood historically, the Front Range suffered. Residents along the area's creeks and rivers, as well as those who lived downstream, were forced to manage surging as well as standing water, particularly in their homes. Almost 600 homes were destroyed or damaged, and more than 900 square miles were affected in Boulder Country. In Larimer County over 1,500 homes were destroyed or damaged, and 1,200 square miles were flooded.[3] While most people did not lose their homes, thousands were forced to clean up flooded basements and crawlspaces. Parking lots full of standing water stranded residents and damaged cars.

A smaller number of people faced more urgent threats. By Saturday, even though the rain had finally relented, rivers and creeks continued to flow at alarming and dangerous rates. Hundreds of residents from Lyons to Longmont and Boulder to Eldorado Springs evacuated battered homes and imperiled communities. The National Guard mobilized an airlift operation, helicoptering as many as a thousand people out of the mountains to safety. Around five hundred fled from Lyons alone. One of the participating guardsmen remarked to a local reporter that he believed it was the largest airlift operation since Hurricane Katrina.[4]

FEW FORCES ARE AS DESTRUCTIVE or as terrifying as surging water. We mostly apprehend the barriers, whether natural or human made, that mark the space between land and water as being impermeable. Rivers and creeks along the Rocky Mountain Front Range, while periodically recognized for their potential to break through containment, are notable as sites where kayakers, tubers, and other water enthusiasts gather. That we regularly see water behaving as we prefer—safely, quietly, curiously but not ominously—trumps our collective sensibility that there is a risk it will not. Visibility and trust in impermeability when it comes to water matters particularly for how we talk about and behave toward what we understand as dangerous.

It is often not until too late that we are forced to reckon with the violence that is always stored in water. Floods are dangerous in part because communities build too close (or are compelled to do so) to riverbanks and oceans. Vulnerable infrastructure and poorly managed towns and cities, many in the United States that are a century or more old and thus paralyzed by the forces of inertia, tempt danger. Citizens, corporate interests, and political officials too often fail to adequately prepare for even well-known risks, including rising rivers and oceanic storm surges near densely populated cities, because it is easy to do so. Socially and economically disadvantaged groups—those discriminated against, pressed into the most vulnerable circumstances, and least able to do much about it— bear the most significant risks, as was the case in New Orleans in 2005. This is so for many reasons, including the pursuit of profit and convenience. The spirit of capitalism, the pursuit of permanent growth and short-term gain are at the heart of the risk, but it is also deeply cultural. There is a powerfully rooted

Faustian sensibility that we are duty-bound to confront and tame nature, particularly in the American West.

Fears about water's power manifest most clearly in moments of urgency, when water trespasses its supposed boundaries and demands immediate attention. Once unbound, water moves according to powers that seem frighteningly its own, although floods are also determined by the physical environment—with the contours of roadways, hills, and footpaths and the intersection of structures and open spaces all directing movement, pace, and violence. Millions of gallons of churning water are made even more terrifying as they uproot and propel collected objects, which exert their own power to destroy. What matters most for those directly affected is the suddenness, the unpredictability, and the overwhelming power of water and the damage it can do. Hurricanes and tsunamis, vastly more destructive than the floods that scarred the Front Range, evoke similar kinds of reckoning with the power of water, whether wind-driven torrential rain or storm surges that have rocked places like Louisiana, New Jersey, and Japan.

Experiencing water as dangerous during floods, hurricanes, storms, and surges also shapes what counts as a crisis, a disaster, or an emergency. Just as important, it also determines what does not. Without suggesting that widespread or large-scale destruction, whether or not anticipated, does not amount to catastrophe, it is noteworthy that what counts as a disaster or as dangerous is determined by its suddenness. There are exceptions, of course, such as in periods of drought or when water resources are dangerously diminished, as in California over the past few years. But when it comes to threats posed by water itself, crisis talk is reserved for tempests and the calamitous. And too often our collective sense of water as dangerous is limited to what we perceive to be natural threats, which discourages critical reflection on the ways that we humans have turned water against us. In the case of natural disasters, scale matters, although there is no precise measure for how large a natural disaster has to be to register as an emergency. Determinations of emergency like the one made by Governor Hickenlooper are appropriate, but they are also limited.

Perhaps most important, emergencies are also meant to be transitional. They are moments through which to persevere, to survive, and ultimately to pass. Environmental catastrophes, including their effect on public health, are not usually understood as permanent conditions. According to the ways that authorities and industry anticipate and subsequently make sense of them, they are fleeting. This image of crisis as temporary, however, comes at considerable cost. There are consequences, especially from dangers that are not seen or immediately felt but are no less potentially catastrophic to the environment and public health.

Whatever our understanding of water's power when pressed by floods and storm surges, collective worry seems to fade as water recedes and reclamation and rebuilding begin. This was certainly the case in Colorado following the flood of 2013, where residents set about the tedious task of community repair in the months and year that followed. The flood has not been forgotten, particularly by those who lost loved ones or homes. But residents along the Front Range set themselves to work in restoring the damage that nature had wrought—a collective community effort that we might usefully perceive as a kind of reconquest.

In moving on, however, little attention has been paid to the ensuing threats—chemical, microscopic, and thus unseen—that Colorado's floodwaters stirred up. These are harbingers of slow environmental violence, and they are every bit as dangerous as the fast-moving water that bore them. They settle into landscapes and bodies and into groundwater. Some of the most dangerous chemical threats resulted from a convergence possible only in places like Colorado, as the floods mixed with chemically laced industrial water used to smash apart shale formations deep below the Front Range. As if dangerous floodwaters were not sufficiently terrifying, the 2013 flood unleashed new threats, including another kind of dangerous water peculiar to and the result of the Front Range's modern energy boom. Indeed particularly when it comes to the convergence of energy and water, we must expand our understanding of what kinds of threats constitute crises.

The dangers from and those that lurked in Colorado's floods are not as evident or as easy to repair as broken homes and roads. These threats have yet to be reckoned with. Because of Colorado's long embrace of extractive industries and because of where these industries have done their work—mining for coal and other minerals historically, and drilling for oil and natural gas today—the flood of 2013 also spilled disturbing amounts of toxins, industrial chemicals, and oil into Front Range waterways and the surrounding environment.

Invisible threats often lurk in floodwaters, although what happened in Colorado, if not unprecedented, has a particular origin. Within days of the start of the Front Range's floods national and local public health authorities, from the Federal Emergency Management Agency to state and municipal offices, warned that floodwater might also be carrying microscopic biological threats. Joel Dyer at the *Boulder Weekly*, citing guidelines from the Occupational Safety and Health Administration, cautioned readers that "floodwater often contains infectious organisms" that can cause typhoid, tetanus, and other diseases.[5] In the end, there were no widespread reported incidents of illness in Colorado, at least not from exposure to microbes, which suggests they were not there. What was there should have been equally alarming. It was not.

Over 2,500 oil and gas wells were shut-in in anticipation of the rising floodwater, ostensibly secured against the threat of leakage, but millions of gallons of water surged over and through many of the twenty thousand active wells in Boulder, Larimer, and Weld counties. While industry efforts to guard against possible environmental damage helped prevent worse, thousands of gallons of pollutants and energy-industry waste were nevertheless washed away, settling into the Front Range countryside and probably into the bodies of some its residents. In his declaration of emergency Governor Hickenlooper acknowledged that among other "known consequences," there was "damage to a natural gas distribution pipeline," although he did not reveal the location, nature, or scale of the damage.

In the months after the floodwaters receded, the Colorado Oil and Gas Conservation Commission (COGCC), the state agency tasked with overseeing and regulating the energy industry, reported that more than 1,600 wells fell within the "flood-impact zone—the area most affected by rushing flood waters."[6] The agency disclosed that it had received fifty spill reports, fourteen of which were "notable," a designation for spills of oil or condensate greater than 20 barrels, the equivalent of over 800 gallons; the largest spill was 323 barrels. In total, COGCC reported, 1,149 barrels of oil and condensate as well as over 1,000 barrels of produced water, chemically laced water injected into wells in the process called hydraulic fracturing, were spilled during the floods. Over 91,000 gallons of gas, oil, and industrial chemicals, including known toxins and industrial wastewater, leaked from the well sites around the Front Range.[7]

FRACKING FOR GAS AND OIL has become increasingly common across the United States, including in Montana, Wyoming, North Dakota, New Mexico, Pennsylvania, Ohio, California, and Colorado. Most of the oil and gas now being drilled sits thousands of feet below the surface, historically considered too expensive to access and too difficult to extract. The recent energy boom is the result of several things. The aggressive

Figure 3. Floodwaters damaged a number of oil and gas sites in Colorado in 2013. Photograph by Bob Pearson/Greenpeace.

exploration for deep oil and gas has been made possible and profitable by consistently high oil prices over the past decade or so; thus the high cost of deep drilling is no longer a meaningful obstacle. Fracking has further expanded the rate and scale of extraction. Where oil wells historically were drilled vertically, today most wells have a single shaft with branches at various depths spread out horizontally. Once the bore well and the extending tentacles have been drilled, they are fractured, or fracked, a process that involves pumping highly pressurized produced water into the well. The combined power of high volume, high pressure, and chemical additives pulverizes the rock, sediment, or shale formations that over time have trapped millions of barrels of oil or cubic feet of gas. Once the oil- and gas-bearing subterranean structures are destroyed, the carbon is freed, able to exit the well, and ultimately transported to refineries and markets.

There are powerful parallels between flooding and fracking. Up to four million gallons of water are used each time a well is fracked. It is an astonishing amount. Already highly pressurized, this large volume of water increases in pressure as it moves underground. Like a flood, as it smashes the rock or shale, the debris adds to its destructive power. Unlike in floods, however, fracking's violence, carried out deep underground, is unseen.

The energy industry's above-ground infrastructure, in contrast, is visible throughout the West. Along Interstate 25 north of Denver and on flat and open county roads through parts of Boulder and Weld counties toward Wyoming are thousands of drilling rigs and tall walls built around well sites to deflect the sound of drilling. Because many operate twenty-four hours a day, they are often lit up at night, illuminating the expanse of territory that the energy industry has come to dominate along the Front Range.

The visibility of the energy industry's surface infrastructure masks what happens underground. There the network of tunnels and wells crisscrosses a vast area, spreading out and running through thousands of square miles, most of it unmapped and unknown to those outside the energy industry. Beyond what we know generally and theoretically, what water does down below is mostly a mystery. This uncertainty is reinforced by the energy industry, which, enabled by state and federal authorities, refuses demands to be more accountable and transparent about what happens underground.

After its use, produced water must also find its way out. Because it is highly pressurized, fracking waste rushes quickly back to the surface. Techniques for the recovery and storage of produced water vary; energy companies either store the waste in open-pit reservoirs on well sites or contain it in sealed, and thus more protected, drums to await transport for disposal. Methods for discarding produced water also vary widely.

Water used in fracking in Pennsylvania is redirected to disposal or water treatment facilities, often in other states because the waste is considered too dangerous to handle locally. In much of the American West, where more space is available, produced water is injected back into the ground, presumably buried in permanent subterranean stasis.

What does it mean to produce water? In order to assure that the injection of pressurized water will achieve the energy industry's aims most effectively, including keeping mud from blocking the well, mere water is not sufficient. Added to it are a cocktail of chemicals and sand, all of which, the companies argue, are necessary for the successful fracturing of geologic formations, for smoothing out friction and potential blockages, and for maximizing impact. There is considerable controversy about the use of chemically laced water to loosen deep oil and gas deposits. Much is unknown, including the impact of subterranean fracking on the environment and especially on fresh groundwater. Most of the uncertainty comes from the lack of transparency on the part of the energy industry. Thanks to a favorable regulatory system that puts minimal demands on them, oil and gas companies have not been forced to disclose the full range of chemicals used in produced water. The federal Energy Policy Act of 2005, steered through by Vice President Dick Cheney, a former chairman and CEO of Halliburton, an oil services company, reexempted the energy industry from a handful of critical environmental safeguards with regard to water, including the Safe Water Drinking Act. Some companies have disclosed these voluntarily and as a result of public pressure. The website FracFocus.org maintains a registry of chemicals that the industry has disclosed publicly. Among hundreds of others, these include methanol (used in antifreeze, paint solvent, and fuel), benzene (a known carcinogen), diesel, lead, hydrogen fluoride, hydrochloric acid, sulfuric acid, and formaldehyde.

In Colorado officials have been unable or unwilling to directly regulate the industry's use of chemicals. Whether or not officials take concerns about the use of chemicals or the industry-friendly regulatory framework seriously is unclear. In any case, the officials do not seem overly concerned. David Neslin, head of the COGCC, remarked in 2011 that his office "encourages oil and gas operators to voluntarily provide information" on which chemicals they use. The lack of political authority and will to push for greater transparency is alarming. Encouragement is not a policy. To be fair, Colorado has mandated a number of requirements, including casing precautions in well bores and storage pits to protect against leakage. For the most part, however, the regulatory system is minimalist. Where we might expect that state officials and regulators would demand that energy companies default toward the protection of the environment and health, the opposite is true. Particularly absurd is the assurance of industrial privilege over community safety. The state demands that fracking chemicals be disclosed to health care professionals, presumably in cases of exposure, although doctors and others treating illnesses must sign confidentiality agreements. Neslin noted, "This allows government officials and medical professionals to investigate and address allegations of chemical contamination . . . while protecting proprietary information."[8] Under the cloak of industrial privacy, energy companies thus have no duty to protect the environments and communities in which they operate.

Oil and gas are not the only things being brought to the surface. Dangerous radioactive sludge, naturally occurring but freed from its containment, has been reported in Pennsylvania and Colorado. Fracking wastewater that has been reinjected in Oklahoma, Colorado, and New Mexico is increasingly linked to an intense and broad pattern of unusual seismic activity, most notably frequent earthquakes. Despite this, and despite the apparent dangers when toxic water is brought into

toxic hormone disruptors—particularly threatening to human, animal, and environmental health—in the Colorado River, which Sandra Postel at *National Geographic* observes is a source of drinking water for thirty million people.[13] The research team determined that of the hundreds of chemicals used in fracking, at least eleven common ones were "compounds that can affect the human hormonal system and have been linked to cancer, birth defects and infertility."[14] Significant chemical traces were found in the Colorado River near a site in Garfield Country, near where a spill of fracking fluids had occurred.

In these cases documented contamination was not the result of what happens underground but of spillage and leaks, common occurrences at drill sites everywhere given the volume of water and the challenge of storing it. Even those who admit that poor water management above ground is sometimes a problem bristle at claims that there is contamination underground, suggesting that because groundwater and oil and gas occupy different and "impermeable" geological strata it is impossible for them to mix. While these claims are seemingly rooted in scientific evidence, they still rely, like the perception of riverbanks as trusted boundaries, on assumptions about permeability and impermeability. One of the central claims made by the energy industry and supportive political officials is that oil and gas drilling and injection practices do not threaten fresh groundwater. Because oil and gas deposits are deep underground, much deeper than most water aquifers, energy experts have consistently declared that rock and other geological formations separate oil far enough from water that no contamination threat exists. But this claim is unproven. Indeed that authorities do not appear to default to a more cautious set of assumptions about impermeability is remarkable, considering that fracking is meant to create permeability and to unsettle the underground in the first place. Neslin said as much before the U.S. Senate: "Thousands of new wells that will be drilled in the coming years rely on hydraulic fracturing to create the permeability that allows fluid and gas to flow into the wellbore and be produced."

There are additional reasons to be wary of such claims, or at least to approach them critically. First, the proof of such claims is deep underground. Given the importance of visibility to how we think about what kind of water counts as dangerous, the potential that invisible chemicals from industrial water are mixing with water requires more urgent scrutiny. The reality, which I explore in more detail below, is that while states like California have disclosed some contamination, in most places it is the energy companies that are solely responsible for tracking and disclosing information about contamination. National and local regulatory burdens, including the 2005 Energy Policy Act that allows energy companies to drill in freshwater formations, are hardly sufficient given the scale of industrial practice.

A second concern is that there is some confusion about how to think about fracking as a system. Drilling for oil and gas may mostly occur well below groundwater, but the reinjection of fracking waste back underground, it turns out, is not sufficiently careful. Among the documented effects are an alarming rate of unusual seismic activity in places like Oklahoma and southern Colorado, where wastewater injection has been linked to hundreds of earthquakes over the past few years. Another threat is systemic pollution, which has intensified in frequency and scale. Among the most dramatic instances are in California, a state that is suffering from multiple environmental pressures. Freshwater is increasingly scarce due to a prolonged drought and wasteful consumption. But California is also home to significant energy reserves and water-intensive fracking practices. The conflict between water scarcity and the heavy demands for personal and agricultural consumption and from the energy industry are unsustainable. Worse, the energy industry's negligent handling of the water it does use has created additional pressure on limited supplies and is raising new threats to the environment and public health. In October 2014 the Center for Biological Diversity, an environmental nonprofit organization, obtained and released records from the state of California that documented massive levels of water contamination from injection wells used to store fracking wastewater underground. The Center reported that state testing of freshwater supplies identified traces of arsenic and thallium, "commonly used in rat poison," near injection wells.[15] Up to three billion gallons of leaking industrial water poisoned aquifers used for drinking and irrigation across central California.

Perhaps the most powerful reason for caution is that the energy companies' promise that it is safely handling freshwater requires trust, trust that has not been earned. Energy companies have made huge efforts since the 1970s to tout their careful stewardship of the environment, widely known as "greenwashing." Such claims are belied by a consistent pattern of environmental negligence, from the *Exxon Valdez* oil spill disaster in Alaska in 1989 to the BP Deepwater Horizon disaster in the Gulf of Mexico in 2010. There are equally significant environmental catastrophes that occur across the Global South, in Nigeria, in Iraq, and elsewhere, that almost never receive similar attention. While disasters on this scale in the United States have captured public attention, there are thousands of smaller and less visible patterns of negligence that happen every day and that the industry is not always forthright about reporting. In some instances, at least, energy practices have prioritized risky practices over environmental safety and public health, including drilling into groundwater.

At the August 2014 American Chemical Society conference in San Francisco Dominic DiGiulio and Robert Jackson of Stanford University reported that "thousands of diesel fuel and millions of gallons of fluids containing numerous inorganic and organic additives were injected directly into" formations in Wyoming that contain both natural gas and sources of drinking water.[16] Neela Banerjee, who reported on the research in the *Los Angeles Times,* cautioned that the two researchers were careful to note that they did not detect evidence of water contamination. Critically, however, they did state that the absence of toxic exposure is hard to determine because the industry makes such scrutiny difficult: "The extent of consequences of these activities are poorly documented, hindering assessments of potential resource damage and human exposure." The bottom line is that energy companies are not committed to environmental protection or public health. They are committed to maximizing profit.

Threats both to water and from toxic water exist alongside other dangers. Most ominous is evidence of toxic chemicals from wells and from wastewater becoming airborne. David Carpenter, director of the Institute for Health and Environment at the State University of New York in Albany, published a study in October 2014 in *Environmental Health* documenting carcinogenic threats lurking in and around fracking sites across the country, including in Colorado. The Southwest Pennsylvania Environmental Health Project, which has been tracking air quality near fracking sites, has documented short but dramatic spikes in dangerous particulates inside people's homes. There is also significant evidence that methane and gas emissions are leaking in greater quantities from wellheads than previously thought. Cornell University–based scientists Robert Howarth, Anthony Ingraffea, and Renee Santoro have estimated that up to 8 percent of the natural gas extracted from a single fracking well leaks into the atmosphere. The concern is methane extracted from shale, which is harmful if inhaled and also a particularly potent greenhouse gas. In fact, according to the Cornell researchers, it has 105 times more "warming impact" than carbon dioxide, making it more dangerous to global climate change than coal or oil.[17]

Scientists and environmental groups have begun to accumulate a significant amount of data demonstrating that the energy industry's mostly unregulated practices are directly threatening people's health. The dangers are no longer theoretical or speculative. Illness from exposure to fracking is tricky to prove, however. Illnesses related to toxic substances, particularly cancer and related diseases, develop over long periods of time and do not require repeated exposure, although regular contact with toxins will greatly increase the likelihood of getting sick. Even small doses, especially from exposure to endocrine disruptors like those found in the Colorado River, can do massive damage over time. Slow-moving and slow-developing diseases mark a kind of environmental violence, but their origins are difficult to pinpoint. Exposure to environmental contamination from toxins used in fracking will almost certainly yield illness, but so will other factors, including lifestyle.

Still, there is compelling evidence that residents who live close to fracking sites, even as far away as a mile, or downstream from them are more likely than others to suffer from air pollution, surface leaks, or hydrological and other contaminations that take place underground.[18] Robert Neuhauser at *U.S. News and World Report* has written on the anxieties and illnesses that accompany proximity to fracking and the toxins it relies on as well as the uncertainty that results from the absence of industry or state efforts to link industrial practice to public health. He reports, "It's difficult to determine which health issues are a result of oil and gas operations and which stem from other factors, because symptoms often start only gradually and government [not industrial] air quality studies have proved limited in scope." He cites Deb Thomas, a Wyoming resident who lives across from a fracking site and participated in a University of Washington study on air pollution, which captures the scale of what are overlapping concerns: "It's very scary. It's very hard to get information. . . . One minute you're living your normal life, the next, people start to get really sick and they can't get any answers."[19]

Symptoms and afflictions range from skin and eye irritations to more serious problems like asthma and cancer. As is the case with toxic exposure, it appears that the environmental and cellular damage is being passed on generationally. A January 2014 study published by scholars from the Colorado School of Public Health and Brown University found a correlation between birth defects and proximity to gas and oil wells. They documented that children of mothers who lived within a ten-mile radius of one of Colorado's fifty thousand oil and gas wells had a 30 percent greater likelihood of being born with congenital heart defects. They also found higher rates of neurological disorders.[20] Pro-energy representatives in Colorado have accused the researchers of using flawed methods and reporting unsupported conclusions. Doug Flanders, the spokesperson for the trade organization Colorado Oil and Gas Association, criticized vaguely that the report had "many deficiencies."[21] It is notable that Colorado energy advocates, rather than citing the preliminary findings as a reason to investigate more fully, have chosen instead to defend industrial practice.

IN SPITE OF A CLEAR PATTERN of pollution, illness, and public anxiety nationally about the dangers of fracking, neither the state of Colorado nor the energy industry was moved to pursue in any meaningful way the threat posed by the combination of the 2013 flood and fracking. Instead state officials turned their attention to better management practices to avoid future floods. Their focus on natural damage rather than industrial pollution should not be surprising. It has long been the case in the United States that industrial practice does not default to being cautious about potential risks to the environment or public health. Why not? There are several reasons, but none that we should accept uncritically.

The first is that political authorities, in Colorado as well as nationally, contest the argument that produced water from the energy industry is dangerous in the first place. The energy industry and its advocates lead the charge in making such denials. The Colorado Oil and Gas Association (COGA), created in 1984, states that its "mission is to foster and promote beneficial, efficient, responsible and environmentally sound development, production and use of Colorado oil and gas reserves." COGA wants consumers to believe that it and the industry take protecting the environment seriously, observing on its website that "our American landscapes and environmental resources are national treasures" and that "safety is the industry's top priority." Such rhetorical posturing is necessary in an era when our collective expectation is, and our need for reassurance demands, that industry will be committed to at least minimal environmental damage. The reality is that the energy industry prefers clever sloganeering and sentimentally crafted public relations language rather than actually implementing robust and expensive safeguards. The expense of more careful environmental behavior, which would cut into corporate bottom lines, is partly deferred to the public via political authorities and national and federal regulatory agencies such as the Environmental Protection Agency.

Just as critical, the industry refuses to concede the real scale of environmental risk. When it comes to toxic threats from fracking water, industry leaders have historically resisted calls that they act transparently in revealing which chemicals they use in the fracking process, citing their industrial and economic right to protect trade secrets. This stubborn refusal to list toxins and the chemical makeup of produced water, however, has come under scrutiny from communities across the country and even from the U.S. Congress, which has directed its research arm to investigate the nature of produced water. Many companies have relented and voluntarily released the information, as noted earlier.

While the energy industry does not deny that some of these chemicals are known carcinogens or are toxic, their defense of corporate practice has been reduced mostly to giving assurances that they are behaving appropriately. Their defenders include politicians. One is Governor Hickenlooper, a former oil company geologist, who is an outspoken advocate for fracking. In 2012 the governor appeared in radio ads paid for by COGA arguing that since passing new regulatory standards in 2008 the state had "not had one instance of groundwater contamination associated with drilling and hydraulic fracturing."[22] COGA states that this was the result of careful technological practice: "The essential factor with respect to the safety of hydraulic fracturing is proper well construction, including casing and cementing to isolate the production formations being stimulated from shallow underground aquifers."[23] While there is plenty of evidence otherwise, from leaks at the wellhead to migrating water underground, the industry insists that its methods are reliable.

Energy companies point to heavy regulatory burdens and the watchful eye of state and federal authorities as constraints on dangerous industry behavior. Yet state and national officials do not, in fact, regulate intensively.

Its advocates have understated the scale of the industry's operations and where its potentially harmful effects are likely to be most severe. COGA asserts, "Drilling is an industrial process and like all forms of energy, creates a footprint on our environment. Drilling for natural gas leaves a small footprint, and most of the infrastructure is underground. Technological advances in the last few years have allowed us to extract 10 times the energy with one-tenth the footprint." At first glance,

Figure 4. Above-ground oil and gas infrastructure dots the Colorado landscape, often hiding what lurks underground. Photograph by Sandy Russell Jones.

COGA seems to be acknowledging the scale of its underground operations. But while it is true that surface operations, thanks to fracking and horizontal drilling, require less space, the scale of the "infrastructure [that] is underground" is what should be of most concern.

The energy industry's defense of itself thus involves the manipulation of the politics of visibility, which I have suggested is central to how we think about some kinds of environmental danger and especially dangerous water. Because fracking happens mostly out of sight and underground, the reliability of the industry's claims that its technological practices—building wells properly, casing them carefully, and ensuring that water formations remain sealed off from potential contamination—are hard to verify. Because of the sheer number of wells, over a million across the country, the reliability of seals and well integrity everywhere is impossible to confirm. This lack of certainty is productive for the industry, which works hard to bury awareness of technological dangers and weaknesses along with its pipelines.

Given the scale and nature of real and potential dangers as well as the energy industry's language in attempting to deflect them, it is possible that the industry is manufacturing doubt deliberately. Perhaps the best-known case is that of the tobacco industry, which peddled the myth of its own innocence and scientific negligence for decades. The truth is the tobacco industry was well aware of the carcinogenic and habit-causing effects of nicotine based on its own laboratory work. The Stanford historian Robert Proctor has shown that Big Tobacco chose not to make that information available, lest it interrupt the pursuit of profit. Cigarette companies also used their wealth to fund alternative science and to support public advocates whose credentials made them appear credible.[24]

In the case of the energy industry I don't have the evidence to prove either the scientific case or that the industry and its political partners are complicit in a massive cover-up. My argument is that evidence is unnecessary because uncertainty suspends both scrutiny and any meaningful political opposition from moving forward.

By defending itself on the one hand while remarking that there is no scientific evidence of danger on the other, the energy industry has offered up a seemingly plausible scientific claim that neither it nor state officials have enough evidence to state definitively that produced water, or that fracking in Colorado and much of the rest of the United States, threatens the American environment or public health in any meaningful way. We should not accept this as a set of scientific claims, the outcome of research or deliberate investigation, but rather as a political tactic. Uncertainty is convenient for industry and for the regulators that defer to it. The result is that the companies dominate what passes for knowledge. Because they own the rights to access and command more resources than either regulatory agencies or environmental or scientific networks that would challenge them, the drilling companies control knowledge. A key consequence of corporate dominance over knowledge about industrial practice is that public political authorities are often put, or choose to be, in a position where they are compelled to use corporate claims as their point of departure. Industrial science and state science have become the same. This is an unhealthy arrangement even without the specter of environmental risk.

If authorities at both the national and local levels, from the offices of the Environmental Protection Agency down to environmental and public health authorities in places like Colorado, Pennsylvania, and elsewhere, are ostensibly serving the interests of citizens, the reality is that they frequently do the opposite. Almost everywhere, regulators mostly privilege the objectives of industry, usually framed through the need to pursue energy security, growth, and jobs—a framework that emphasizes the importance of markets, trade, and the economy—rather than sustainability, health, or the environment. The privileged position of industry in America, in which advocates of economic and political liberalism (support for open markets and free trade rather than protectionism) rarely criticize corporate excess, has a long history.

There was a moment in the early twentieth century when trusts and monopoly capitalism, including the early oil industry, came under fire, but since the 1970s the reconsolidation of wealth in the hands of a few elites and a few corporations has mostly been celebrated rather than criticized. The dismantling of the regulatory state and the freeing of business to pursue its own aims, often heralded as a form of Reaganism but widely embraced across the political spectrum, has been central in America since the closing decades of the twentieth century. The consequence is that regulators have struggled against pressure to look the other way, at best, or concede ground altogether.

This has certainly been true for the domestic energy industry since the early 1970s, which enjoyed considerable room to maneuver in the aftermath of the energy crisis in the middle of the decade. As discussed in chapter 1, the U.S. government, fearful that dependence on foreign oil left the country vulnerable to economic and other threats, freed the energy industry to pursue resources wherever they were available. It took several decades for technology to make the intensive extraction of deep oil and gas possible and cost-effective, but the political framework for the current boom was in place. What is perhaps most remarkable about this historical deference to the energy industry and its dangerous practices is that it emerged at precisely the moment historically when U.S. citizens and political leaders had begun, finally, to think critically and systematically about the need for more robust environmental protections. Recall that the 1970s were the decade in which there was some meaningful movement toward a strong regulatory effort at protecting complex ecological systems. It turned out, however, that the energy industry would come to enjoy significant exemptions. These are critical to understand in order to make sense of why today's energy practices are not understood as dangerous and why they are underregulated.

Institutional and political efforts to protect the environment, including water, emerged fitfully over the course of the twentieth century and in response to increasingly visible problems. The absence of environmental protectionist measures through the 1960s helped birth the modern environmental movement, which began targeting widespread industrial practices that endangered the environment, outlined elegantly and most persuasively by Rachel Carson in her seminal work, *Silent Spring*, in 1962. This development would ultimately lead to two important outcomes. First, it created enough pressure to facilitate the creation of a regulatory system. Second, it helped shape the terms by which the environment and water could and would be known. It turned out that, in spite of all its good intentions, the results proved limited and, although unwittingly, created space for the energy industry to carve out favor and exemptions for its own practices.

Shortly after the creation of the EPA in 1970, Congress passed the Safe Water Drinking Act (SWDA) in 1974, which was subsequently amended in 1986 and 1996. The SWDA aims "to protect public health by regulating the nation's public drinking supply . . . [and by setting] national health-based standards for drinking water to protect against both naturally occurring and man-made contaminants."[25] Through the amended SWDA the EPA not only oversees water treatment but also the protection of water at its source. The SWDA is also charged with the responsibility of regulating the injection of waste into more than 800,000 wells underground. Abrahm Lustgarten, a reporter for the online news site ProPublica, has written that "subterranean waste disposal . . . is a cornerstone of the nation's economy, relied upon by the pharmaceutical, agricultural, and chemical industries" today.[26] The EPA sees underground injection as central to the U.S. economy: "Our way of life would be quite different without injection wells. Agribusiness and the chemical and petroleum industries as we know them today, could not exist" without the practice of storing toxic waste out of sight and underground.[27] Given the pattern of contamination that had begun to emerge in the 1960s, however, the need to regulate had to be squared with the EPA's position that the practice needed to be maintained.

A decade after its creation, the EPA began setting standards for underground waste disposal, for regular inspections of hazardous waste, and for how to think about waste, toxicity, and the meaning of the term hazardous itself. In 1980 the agency created a classification system for underground disposal wells that ranked them according to risk. Some of the risks to be avoided were obvious, including the need to keep sewage, radioactive waste, and other kinds of toxins out of freshwater supplies. Others, including waste from the energy industry, ended up being contested and eventually exempted, to the sole benefit of oil and gas companies.

The mandate to oversee and manage hazardous waste in ways consistent with the SWDA was set in place in 1976, when the Resource Conservation and Recovery Act (RCRA) authorized the EPA to "control hazardous waste from 'cradle-to-grave.'"[28] Almost immediately the energy industry put pressure on the government, insisting that the substantial regulatory demands were too expensive. The EPA yielded ground and eventually moved to exempt the industry from managing the toxic dangers that the RCRA sought to mandate. As Joel Dyer and Jefferson Dodge have written in the *Boulder Weekly*, "The mother of all oil and gas waste exemptions had its beginnings in 1978 when the EPA proposed reduced requirements for a couple of types of large-volume wastes associated with the oil and gas industry, namely produced water and drilling muds."[29] It took the EPA a decade to resolve the matter, finalizing the energy industry's exemption from regulation in 1988. Steered by the antiregulatory ideology of the Reagan administration, the EPA determined that the energy industry's practice of using "produced water, and other wastes associated with the exploration, development, or production of crude oil or natural gas" was not hazardous.[30] As Lustgarten has noted, "All material resulting from the oil and gas drilling process is considered non-hazardous, regardless of its content or toxicity. . . . The new approach removed many of the constraints on the oil and gas industry. They were no longer required to conduct seismic tests. . . . Operators were allowed to test their wells less frequently for mechanical integrity."[31] By the end of the 1980s the rules for environmental protection that were taking shape for other industries did not apply to energy.

The EPA's decision was deeply contradictory. On the third page of the very report in which the agency exempted the energy industry from too cumbersome a regulatory burden regarding hazardous waste, agency officials wrote, "Oil, gas and geothermal wastes originate in very diverse ecologic settings and contain a wide variety of hazardous constituents. EPA documented 62 damage cases resulting from the management of these wastes, but found that many of these were in violation of existing State and Federal requirements."[32] Thus the EPA acknowledged that the energy industry's practices were toxic and hazardous, threatening to both the environment and public health, and yet not worthy of its close scrutiny.

As has often been the case historically, the government's decision had little to do with scientific evidence or the real dangers posed by industrial practice. Instead it was based on cost and a set of assumptions about the centrality of the energy industry to the national economy and to notions about national security, shaped by the energy crisis. The convergence of anxieties around dependence on foreign oil, questions about the need for a regulatory state more generally, and the view that the energy industry was "too big to fail" resulted in regulatory authorities surrendering almost entirely to the energy industry. In a 1988 report the EPA asserted that the regulation requirements outlined by the RCRA of 1976 presented "several serious problems." These included "an unusually large number of highly detailed statutory requirements" as well as offering "too little flexibility to take into account the varying geological, climatological, geographic, and other differences characteristic of oil and gas drilling and production sites across the country." The report further noted that the RCRA did "not provide the Agency with the flexibility to consider costs when applying these requirements to oil and gas wastes."[33]

Regulation of the energy industry, then, was seen as prohibitively costly for the EPA to undertake. It is true that the EPA has always been chronically underfunded and weak politically, but it is crucial to recognize that the energy industry was exempted from regulation for reasons of cost and not because its practices were not in fact dangerous. As Dyer and Jefferson observed:

> Even though this waste is hazardous to human health and the environment, the EPA [exempted] it from federal laws written to protect the public from toxic waste because the problem is so massive . . . that it would make it literally impossible to drill for oil and gas in the U.S. if the industry had to pick up the tab for remediating the contamination it creates. It is just math. More than 10 barrels of waste are created for every barrel of oil. Having to properly deal with this waste . . . would be so cost-prohibitive as to be a threat to our economy, and, therefore, our security.[34]

Failed EPA efforts to follow through on concerns about the environmental costs of fracking raise questions about its own efficacy and to whom it ultimately answers. In 2010 Congress ordered the EPA to investigate the dangers that fracking posed to drinking water, a matter of considerable urgency. Rather than moving quickly, however, the EPA appears to have stonewalled. The research was expected to be completed in 2014, but in the spring of 2013 the EPA said that it was delaying its findings for up to three more years.

In 2011, after three years of study, the EPA released a draft report of research it had done in response to complaints that citizens in the town of Pavilion, Wyoming, were suffering the ill effects of local fracking. The report concluded, "Data suggest that enhanced migration of gas has occurred within ground water at depths used for domestic water supply."[35] In tests of local water the EPA identified the presence of chemical compounds that could only have come from fracking fluids and levels of benzene and methane that far surpassed federal standards. The EPA and its report came under immediate fire from both the energy industry and the state of Wyoming, which generates a tremendous amount of revenue from gas exploration and production. Pro-energy advocates called into question the EPA's science and suggested that the report was spurious. In June 2013, instead of following through, the agency decided not to complete the study or to follow up on its provisional conclusions. Instead it left further research to the state of Wyoming, a clear conflict of interest considering the state's support of the energy industry.

GIVEN THE RECURRING PATTERN of environmental agencies and political officials deferring to the interests of the energy industry, it should not be surprising that the toxic dangers likely unleashed by the Colorado floods of 2013 have been mostly ignored. Critics of the energy industry and its environmental practices often point back to 2005, when the George W. Bush administration pushed through an Energy Act that eliminated the regulatory burden on oil and gas companies, extending the exemption of practices and waste from protecting water. The reality, though, is that the demand for the industry to do so has almost always been nonexistent. The EPA and other federal authorities continue to shift the burden of regulation onto state governments, which have shown little interest in challenging the industry.

For its part, Colorado has appeared to ramp up regulatory efforts, setting higher standards than anywhere else for methane emissions and other controls. But while Colorado's standards and stated aims for confronting toxic oil and gas wells seem reasonable, they fail in practice. This is partly because the state's top officials, including Governor Hickenlooper, are not concerned about the nature of the threat, having advocated openly on behalf of the industry and denying that its practices are toxic. It is also partly because the regulatory standards are not backed up by the necessary administrative capacity to enforce them. This is not because there is no interest among some officials for more regulatory oversight. In 2013 the state legislature passed a bill increasing the number of state inspectors, responsible for monitoring fifty thousand wells, from sixteen to twenty-seven. Reasonable observers, however, would expect the number of inspectors to be many times larger. Hickenlooper vetoed the bill. The governor also cut fines for oil and gas companies found to be at fault for environmental contamination.

Some Colorado officials insist that because the threat to water, the environment, and public health more generally does not exist, additional regulatory requirements are unnecessary. Challenges to these claims are paralyzed by the politics of uncertainty on the one hand and, on the other, the convergence of industrial science with the absence of political will to confront the power of the energy industry.

In the end, what citizens in Colorado and elsewhere are left with is ominous. The scale of the threat is denied by the very forces that should be most committed to confronting it. Instead powerful political leaders have not just conceded to the energy industry; they have sided with it. In the essay that follows, I explore some of the broad political implications of this alliance, including the emergence of antidemocratic politics in energy-rich communities.

When it comes to the matter of what counts as an environmental danger, what we consider to be a crisis, and the very possibility for certain kinds of environmental politics, the unfortunate reality is that thanks to an unhealthy industrial-political alliance, a stronger state regulatory apparatus or, better yet, political will to support the pursuit of alternative energy sources that are not dangerous at all is unavailable. Whatever the prevailing rhetoric, the material reality on the ground and under it is that powerful threats to our ecological order and collective human and nonhuman health are at work.

CHAPTER 3

Arabia on the Front Range

CARBON-BASED ENERGY RESOURCES—oil, natural gas, and coal—fueled the most dramatic transformation in human history in the nineteenth and twentieth centuries, powering the rise of industrial society and the spread of global capitalism. Oil and, more recently, natural gas have been particularly critical as sources of energy in the modern era. From industry to transportation to the ways we live in the world, cheap carbon energy has been *the* world's single greatest transformative force. But after over two centuries of oil's dominance and our reliance on it, it is clear that the romance has been a corrosive affair.

The clearest evidence of damage is carbon energy's devastating environmental impact. Global dependence on oil and gas for industry, especially for transportation, has radically altered the world's climate, mostly as a result of the massive emissions of airborne waste. The planet's skies are filled with warm gases generated by human production and consumption habits. The dangers are hard to overstate. Rising global temperatures threaten a cascade of calamities, from melting icecaps to rising ocean tides and the generation of superstorms and other heavy weather. Perhaps the ultimate tragedy is that climate change, set in motion by the habits of Western society, now threatens everyone.

While this climate change marks a new and perhaps the ultimate danger, carbon energy's pernicious environmental legacy has a long history. From western Pennsylvania in the nineteenth century to the oil fields and refineries of the Global South today, in places like Nigeria and Iraq, routine oil spills have poisoned landscapes and soil and water resources.[1] Aside from industrial accidents and the leaks we are periodically aware of, the very nature of extracting oil and gas from underground has always been a toxic process.

Spills, environmentally dangerous practices, the lack of regulation, and an industry so powerful that it has been able to hide its abuses have made the carbon energy industry among the single most dangerous polluters ever. Efforts to better protect the environment through regulation in the late twentieth century have either failed or been undermined by an industry that arguably possesses as much political power as it does wealth. Consider the aftermath of the massive oil spill from the destruction of British Petroleum's Deepwater Horizon rig in the Gulf of Mexico in 2010. According to the U.S. government, almost five million gallons of oil spilled into the Gulf, where it wrecked whole ecosystems and disrupted economic life along the coast. The oil settled on the floor of the Gulf, where it will continue to mete out environmental harm for generations. BP has been fined billions of dollars for its negligence. Remarkably, though, not only has drilling for oil in the Gulf continued, but the appetite for oil has sparked greater demand that energy companies explore for oil and natural gas in even more precarious places and ways. The Keystone XL pipeline, for instance, would deliver Canadian oil to the Gulf across the American heartland, putting thousands of miles and communities at risk of spills and contamination. It is as though the lessons of the BP disaster or the smaller oil and gas spills that occur every day all over the world never register at all.

Oil and gas have been equally poisonous for politics. Oil's rise as the twentieth century's greatest prize has made it a particularly powerful engine of corporate, political, and imperial greed. Intimacy between commercial and political power is hardly new. The history of industrialization and its global spread, including the European pursuit of resources and markets through empire, is littered with commercial and political alliances. The lines between political and commercial interests are particularly blurred when it comes to control over the environment. The American version of this story, both domestically and globally, follows the imperial blueprint. The drive to harness the power of rivers has long yielded political concessions for businesses from New England to the American West. And the government has encouraged digging for coal, gold, silver, and other minerals by limiting the laws around ownership of the land and what lies underneath it.

In colonial America it was privileged white men. Similar patterns of privilege, often recapitulating racial injustice, took hold in far-flung colonies and in the postcolonial world, with access to and rights to control the environment and the ways it could be put to work and profit tightly controlled by political powers. Water and other sources of power have been sites of struggle in which the differences between economic and political interest have collapsed. Carbon energy, especially oil and its power to mobilize avarice in the past 150 years, has stood apart.

Partly because of greed and partly because oil made new kinds of political opportunity possible, its most important impact is that it has been an antidemocracy machine. The struggle to control oil has often been linked closely to the struggle for democratic rights. At least from the perspective of those who desire political pluralism, wide room for political rights, and participation, oil has been a curse. This is especially true in formerly colonized or developing countries in the Middle East, South America, and Africa, where much of the world's oil resources are located. In many of these places greedy global oil corporations conspired with local elites to undermine the possibility of democratic politics and to build and then protect some of the world's worst tyrannies in the early to mid-twentieth century. Many of these remain in place today. From the view of the oil companies, small, strong, centralized oligarchies were preferable to big, inclusive, nationalist governments that would have prioritized the rights of locals over corporate profit and elite privilege. Imperial and geopolitical power followed the corporations. After World War II U.S. policymakers framed their interests in the oil-rich parts of the world like Saudi Arabia and the Persian Gulf as maintaining security and stability and furthered antidemocratic politics there in order to ensure that friendly regimes remained in power.

Oil giants and Western powers, then, bent small states to their corporate and political will, all in the interest of maintaining access and control. Even after the oil companies left the developing world in the 1960s and 1970s, the tyrants remained, as did Western support for their survival. Great wealth from oil thus came at tremendous political cost.

The antidemocratic impact of oil wealth has been most closely associated with the developing world—the Global South and formerly colonized countries that have struggled to find their way after the end of empire. However, as detailed in chapter 2, the corrosive political impact of oil is not restricted to the non-West. The energy boom in the United States over the past decade is pitting the rights of U.S. citizens against the rights of the energy industry, often with worrisome consequences for democratic politics.

The recent energy boom in the United States marks the rise of a new generation of oil barons, who, like the Carnegies and Rockefellers who shaped American politics a century ago, prioritize their own profit over the health of the environment or local communities. The old oil barons built empires that grew so large, powerful, corrupt, and rapacious that it required a combination of scandal and the massive power of the federal government to dismantle them. Now that Big Oil has risen again in the United States, it is important to publicize its ability to overpower political systems and shape them in ways that suit their business interests.

In states like Colorado and Pennsylvania community efforts to protect the environment and public health and check the power of energy companies and state authorities are under fire. While it is a stretch to argue that Colorado and other energy-rich parts of the United States are following the political path of Saudi Arabia, there are alarming antidemocratic patterns taking hold. It is useful to consider some of these and how they are taking shape. There was nothing inevitable about how authoritarian regimes in the Global South came to be. There is also nothing inevitable about the persistence of democratic politics in America.

Oil's rise to hegemony, including both its industrial utility and the scale of its political reach, has also compelled otherwise critical groups to support poor choices, not necessarily because they are misguided but because the range of options has been undermined by the energy industry itself. Some of the environmental groups that are most critical of our reliance on carbon energy because of its impact on the climate see in the rush to develop natural gas the least-worst carbon energy option available. Compared to coal, which is an especially dirty source of power, natural gas and shale oil have been posited as bridge fuels to an alternative energy future. These are fanciful and mistaken claims that have made for strange and dangerous alliances between ostensible foes. It turns out that fracking is more dangerous to the climate than we have been led to believe. What is more important here, however, is not the details of the scientific argument about which carbon energy resource is most dangerous, but that oil and energy companies have become so powerful, so politically influential, and so difficult to confront that they have successfully narrowed the range of energy options. These are false choices. The political consequence of this dilemma is that we have yet to move beyond carbon at all and that well-intended groups that are environmentally progressive have unwittingly enabled the very environmentally damaging and antidemocratic forces they oppose.

While the energy industry wages a campaign against democracy, against public health, and against the environment, there is also something hopeful about the current moment. The response to the recent energy boom has been mixed, a combination of enthusiasm (Drill, baby, drill!), apathy, and alarm. Many individuals and communities are pushing back against the power and privilege of Big Energy. And they are doing so as both environmental and democratic activists. Given the scale of contemporary challenges—including climate change, the dangers that energy companies pose to local environments everywhere, and corresponding threats to public health—environmentally driven struggles for greater political rights and the struggle to resist the antidemocratic power of Big Energy are urgently necessary.

Before turning to the politics of the energy boom in the United States today, it is worth deliberating on oil's historical relationship to democracy in oil-rich states in the colonial and postcolonial worlds. There are important similarities as well as limits to the comparison. What energy companies have wrought globally is part of the same historical trajectory at work today in the United States. That energy politics in Pennsylvania and Colorado resemble practices in places like Iran and Saudi Arabia should be startling. Yet there is nothing exceptional about the United States or its political order; it too is vulnerable to the power and interests of those who are concerned more with profit than democracy.

Figure 5. Energy companies in Colorado build walls around their well sites to diminish noise pollution. Their efforts to contain environmental contamination are not so visible. Photograph by Sandy Russell Jones.

But there are also important limits, at least for now, to the connections between Big Energy's operations abroad and in the United States. The scale of violence and the absence of democratic rights in energy-rich communities in the Global South are an order of magnitude greater than in the United States. While democratic power in some energy-rich American communities is under siege, the terms of citizenship and the risks of dissent are starkly different and more dangerous in Saudi Arabia, Iran, Iraq, Russia, and Nigeria. Critics of the political system in those countries routinely disappear or are imprisoned. But we should not take for granted that democratic rights will survive without a constant struggle to protect them. While the energy industry does not seek to roll back *all* community rights, who is to say that the erosion of popular authority over the environment, public health, or how industry behaves will not lead to the more systematic chipping away of popular sovereignty? It will hardly be surprising to students of the history of industry that preventing big business from bending political systems and privilege to their advantage requires constant attention.

The large energy companies building favorable political alliances in Pennsylvania, Colorado, and elsewhere were not the first to do so. Mining companies and the railroads in the American West behaved similarly in the past. Extractive industries in the United States have long sought to undermine democratic rights in the pursuit of profit. The large oil companies that struck it rich in places like Saudi Arabia, Iran, and Mexico and that helped hollow out political systems in the Global South in the early to mid-twentieth century were applying abroad practices they had sharpened first in the United States. There were, of course, European forces at work too, particularly from England. They too pursued an antidemocratic course in places like Iran and Iraq. Antidemocratic tools developed in the American South, the American West, and the English countryside proved indispensable to companies like Standard Oil of California (now Chevron), which helped build a Jim Crow–style system of labor in Saudi Arabia, in which Arabs and nonwhite workers faced systemic racial discrimination and were subjected to wretched living and working conditions. To understand how far energy companies are willing to go politically and environmentally in extracting oil and gas and maximum profit, then, it is important to understand their patterns of behavior and their record.

In his provocative recent work on "carbon democracy," the political scientist Timothy Mitchell argues that struggles to control carbon-based energy resources have been central to politics and to the possibility of democracy since the nineteenth century.[2] It turns out that not all energy resources lead to similar kinds of politics, a fact that has much to do with the resources' physical qualities as well as where they are located. Until the late nineteenth century, the machines and tools that powered the rise of modern industry and capitalism were fueled by coal. Coal remains important to industry and especially to the production of electricity today, but before the rise of oil it was unrivaled as *the* essential source of power. In both the United States and large parts of Europe, especially England, factories and centers of industrial production were often located close to coal mines. Because mining for coal required massive human labor and because it was heavy and difficult to transport, the proximity of coal mines to centers of industry made sense.

Coal's global hegemony came to an end in the late nineteenth century with the rise of oil. The transition from coal to oil is often described as the result of technological progress or "the market." Oil is lighter and easier to manage, and because of its physical qualities it required less labor and human interaction to extract from underground. The most intensive phase of oil labor was exploration and drilling. Once oil wells had been put in place, what was most needed for a successful operation was money to finance transportation and refining. Compared to coal, oil was capital-intensive rather than labor-intensive. It has long been the case that capitalists prefer cheaper machine labor to expensive human labor, and the early energy industry was no different. On the consumption side, oil is easier to use. With the rise of the internal combustion engine, fueled by refined petroleum products like diesel and gasoline, oil was more efficient and cleaner than coal. Its use could be adapted to smaller machines, enabling the production of smaller and faster or larger and more heavily armed naval vessels, automobiles, and eventually other modes of transport. Refined oil was also better suited to heating homes. Wealthy investors thus saw greater commercial opportunity in oil and capitalized on its appeal as a better source of energy and fuel for modern society. Alongside technological explanations, then, are claims that the turn to oil made sense given the scale of profit that was possible from controlling its extraction and distribution. This is the way markets function, we are told: cost, supply, and demand drive change.

These are compelling narratives, mostly because it is true that oil was a superior source of fuel and power and was cheaper to produce and more profitable. But there are limits to these explanations for its rise to dominance. As a commodity with certain kinds of political power, oil was especially appealing to political elites because of its antidemocratic qualities. Because coal was so labor-intensive, coal mines and the networks in which coal moved and was used became centers of labor organization, unrest, and spectacular violence. In the nineteenth century in England and in places like Colorado (as shown brilliantly by the historian Thomas Andrews in his work *Killing for Coal*) and Pennsylvania, radical labor activism and deadly standoffs between workers and political and business elites were especially pronounced at coal mines.[3] This was not good for business or for the various ways political and corporate elites sought to carve out space for their own privilege.

Strong labor and union activism in Western coal communities could disrupt production and supply. Demanding better wages, better care, and less predatory behavior from mine owners and factory bosses, workers threatened profits by striking, destroying equipment, and paralyzing extraction. Because they possessed the power of their labor, workers directly threatened bourgeois interests. Coal miners threatened more than just the economic interests of commercial elites, however. They also threatened relations between state and corporate interests. Insisting on greater rights at work, miners also pressed political authorities for a more favorable regulatory system to protect them.

Elites had little interest in yielding to such demands. One result was political support for the pursuit of alternative sources of energy. A particularly clear example of this was in England. Historians of oil and energy explain that in the twentieth century the British decided to convert their coal-based navy to oil, which allowed them to build more efficient, larger, and faster ships. The transformation required a source of energy abroad, since they had yet to discover oil closer to home. That source ended up being in the Middle East, in what was then Persia, and then in larger quantities in Iraq, which the British occupied during World War I. The British obsession with the Middle East and especially with the oil in the Persian Gulf would endure until the middle of the century.

However, claims that this obsession was based entirely on strategic interest miss an important part of the political story. As Tim Mitchell shows, British leaders were more concerned about striking English coal miners and their democratic demands than they were with technological innovation. The British Navy's desire for Persian oil, then, was as much about stamping out democracy at home as it was about finding much needed strategic energy resources abroad.

Oil's antidemocratic power in England and the United States built an empire of energy. In order to secure control over far-flung oil resources, Western powers and the giant American and European oil companies scrambled for access and then dominance in Mexico, Iraq, Iran, and Saudi Arabia. The drive for oil extended into parts of Africa and Southeast Asia as well, where Western corporate dominance remains today, even if formal empire has receded. Oil-rich states in the colonial and then the postcolonial world have routinely failed to become democratic. Instead they are mostly authoritarian, oppressive, and sometimes violent. Power was historically consolidated in centralized government institutions and monopolized by small networks of political elites and oligarchs. This did not happen on its own; it required constant intervention and the flexing of both corporate and political will.

Reflecting on the long steady march of antidemocratic politics in the world's most important oil-producing states, observers and critics, including citizens of those states, argue that oil is a curse. They often mean this in two ways. First, oil and gas producers have mostly failed to develop alternative industries, leaving them economically vulnerable to price shocks and market uncertainties. Second, energy wealth has hollowed out political systems in oil-rich states, allowing for the rise of small groups of powerful elites that jealously hoard power. The idea of oil as a curse is certainly not news to those closest to it. In 1976 Juan Pablo Pérez Alfonzo, a former oil official in Venezuela and one of the cofounders of OPEC, remarked, "Oil will bring us ruin. . . . It is the Devil's excrement."[4] Ahmed Yamani, Alfonzo's counterpart in Saudi Arabia, the world's most important oil producer, is said to have remarked, "All in all, I wish we had found water."[5] These sentiments are certainly shared by those on the outside of the political system looking in, who would trade oil wealth for political rights.

Over the past several decades observers who have sought to make sense of the absence of democracy in energy states have argued that political inequality in those places is the result of the fantastic wealth generated by oil. Money, lots of it, from the sale of oil has accumulated in banks and treasuries dominated by political elites. In seeking to maximize their own power and privilege, rulers and leaders in oil states are said to have struck a bargain with those over whom they rule. In exchange for their lack of political power, including the right to organize, assemble, and speak freely, citizens are well cared for by welfare states that redistribute oil wealth through cradle-to-grave services, subsidies on food and water, and free entertainment. There is some truth to this. Political leaders in energy-rich communities built massive redistributive states, using oil wealth as a form of patronage in order to pacify citizens and buy off dissent. They have touted energy as an engine of opportunity, providing jobs, security, economic stability, and so on. These have often proven illusory, however, and citizens from Saudi Arabia to Iran have routinely bristled against the authoritarianism.

Yet the focus on wealth to explain authoritarianism's origins in energy-rich countries obscures more than it reveals. Yes, the massive revenues from oil fund petroleum-fueled autocrats. But wealth alone does not explain the lack of democracy in these states. A better explanation is that large Western oil companies, often backed by military power, helped build, shape, and support the rise of autocracies instead of democracies. They did so deliberately.

As was the case closer to home, the way that oil was extracted, with limited labor and thanks almost entirely to the capital investment and initiative of Western backers, allowed corporate elites to partner with small groups of allies to build centralized and modern states. Commercial and political power and interest were indistinguishable. Democracy could not be tolerated in the oil-rich parts of the empire or even after decolonization, as democrats who come to power tend to favor the interests of their fellow citizens rather than foreign business and consumers who want cheap gasoline. Antidemocracy in oil-rich states was established in the first half of the twentieth century and has survived ever since, although not without challenge. Indeed with internal and regional threats in the Middle East, protecting the autocrats has required almost constant intervention, sometimes violent, by the great powers and their local allies.

Both the oil companies and the rulers of Saudi Arabia counter that their efforts in extracting and selling oil were undertaken with an interest in developing and modernizing the country. A powerful narrative persists that the Arabian Peninsula was trapped in the seventh century and that without oil and the efforts of the oil companies, it would have remained so. It is true that oil and the wealth it brought led to massive transformations, but the politics of modernization, of "improving" states like Saudi Arabia, often had either the intent or the effect of producing illiberal outcomes as well, including an oppressive political system. In addition, in ways that are reflected in the politics at the heart of the struggle in places like Colorado today, the way autocratic systems were built in the energy-rich parts of the Global South was often at the expense of the environment and the ecologies that bound humans with the nature around them. In addition to limiting political rights, then, the making of modern states and systems in Saudi Arabia and similar oil outposts resulted in massive environmental harm.

The corporate conglomerate that operated in Saudi Arabia from the 1930s to the early 1980s, the Arabian American Oil Company (ARAMCO), was an energy giant on the Arabian Peninsula and is still the largest and arguably the most important oil producer in the world. In pursuing and profiting from Saudi Arabian oil, ARAMCO destroyed democratic possibilities by aligning with autocratic elements of the Saudi royal family against more liberal factions in the 1950s and in waging a tireless campaign against organized labor, the country's most outspoken force for democratic possibility, in the 1940s and 1950s.[6] At every step ARAMCO struggled against the social and political forces in Arabia that sought a more inclusive political system.

It did so by developing the country's physical environment, in the process helping to construct centralized political agencies. Even though Saudi Arabia had a small population for much of the twentieth century, its rulers, aspiring to models of modern governance and modern statehood, understood that controlling their citizens required establishing power over their bodies, the land they resided and toiled on, and especially their need for water. After all, Saudi Arabia has no natural lakes or rivers, so water, essential to life, is a tremendous potential source of state power.[7] ARAMCO helped the Saudis shore up control over water by lending its engineering and technological strength to developmental and environmental projects that seemed to be solving the water scarcity issue but that were really instruments that brought the state into communities that did not necessarily want it there. Water's centrality to Saudi Arabia's political authority remains in place today. The kingdom continues to heavily subsidize water for both industrial and personal consumption, even though doing so is expensive and requires the use of its own energy resources. Saudi Arabia must use massive amounts of its oil and gas to fire the desalination plants that make freshwater available, oil and gas that might otherwise be sold globally.

The alliance between ARAMCO and the Saudi state also produced terrible environmental consequences, including the destruction of important water resources. In a small oasis known as Al-Hasa that is close to Saudi Arabia's Persian Gulf coast and home to rich resources of both oil and water, it is widely believed that ARAMCO stole water from local farmers in order to help expand the production of oil from the famous Ghawar oil field, the largest ever discovered. Oil companies have long used water to help bring oil to the surface, injecting it into wells in order to create pressure to push oil upward. ARAMCO relied on this technique at Ghawar, which is adjacent to an old agriculturally and water-rich community of mostly date farmers. In doing so, though, ARAMCO, probably with the knowledge of Saudi authorities, took the water from underground resources that had historically served to sustain regional agriculture. Because it enthusiastically and wastefully used massive quantities of water to help extract oil, ARAMCO ended up draining the resource that local farmers had relied on for generations, killing off local agriculture. The community of Al-Hasa has never recovered.

It was not the case, then, that democracy in places like Saudi Arabia failed because oil wealth made it impossible.

Democracy failed because the conditions of its possibility were deliberately destroyed, often through a combination of practices, some of the most fundamental being the manipulation of energy and other environmental resources. It is also important to recognize the scale of the effort that was required to establish authoritarian states. Massive cost, intervention, and an ongoing commitment were necessary. Authoritarianism and the absence of democracy in Saudi Arabia were not necessarily the conditions that prevailed when ARAMCO began its work in the 1930s, but they were the result.

Thus democracy in the oil-rich parts of the Global South was *unmade*. While the depth of autocratic power and the oppressive and violent character of places like Saudi Arabia and Iraq mark extreme cases and do not map neatly onto the American West, the broad lessons—that these systems were crafted by a small group of political and corporate elites— are applicable today to newly emerging centers of energy in the United States. During the most recent energy boom, it has become clear that the energy industry and its allied local political partners are systematically chipping away at the democratic power of the communities in which they operate. It is not yet reasonable to suggest that Colorado or Pennsylvania is becoming a new Arabia, but it is worth drawing attention to the parallels between them. Unlike the oil-rich states where dissent and challenges to dominance have been criminalized and where critics of political excess suffer terribly, there are resilient opponents to what is happening in oil-and-gas-rich communities across the United States. They are under tremendous pressure, however.

THE SHALE AND NATURAL GAS energy revolution in the United States has yielded spectacular results, with over a million fracking wells across the United States. Production levels and estimates of just how much oil and natural gas sit in formations across the country have grown greatly in the past ten years. In 2012 the U.S. Energy Information Administration estimated that in American shale plays there are almost 500 trillion cubic feet of gas. Across the new energy landscape landowners and residents, not to mention a new generation of oil-field workers and drillers who have hurried to take advantage of short-lived but high-paying jobs, have embraced the frenetic rush.

There are powerful incentives for them to do so. While oil field labor is mostly necessary only during the exploratory and drilling stages of the production cycle, wages rank among the most lucrative. Considering the devastating impact of the financial collapse in 2008 and a new American economy that relies less on labor generally, the energy industry offers an income that is increasingly rare. The result is that thousands of workers from across the United States have converged on some of the most remote parts of the country, places where drilling and exploring are most intense. Across the American West and Southwest, including New Mexico, Colorado, Montana, and North Dakota, the effects are evident. Oil facilities are everywhere, as are the work trucks, the work crews, and related evidence that the energy industry's success is materially rewarding for a few. Hotels are full of temporary workers who move from community to community. They operate drilling equipment, build infrastructure, and drive trucks that carry tools, water, and waste. There are distributed opportunities as well, including in places like Wisconsin, where workers extract from wastewater the sand that is used in the fracturing process. The drop in oil prices over the second half of 2014 and in early 2015 did lead to some decline in work opportunity, although it remains to be seen whether the energy industry's operations will be significantly diminished.

Thousands of landowners who own their own mineral rights have responded similarly. With the possibility that oil or gas, engines of spectacular wealth, sits below their feet, many have leased or sold their rights to energy companies in the hope that they will strike black gold. My own family has leased its mineral rights in southwest Mississippi to energy companies hard at work extracting oil from the Tuscaloosa Shale Formation. The incentive to do so does not come down to greed, at least not most of the time. In Pennsylvania, home to the Marcellus Shale Formation, and in Weld County, Colorado, the decision to support the energy industry often provides financial security or opportunity where other possibilities for income have dried up. The reality in the United States is that anxieties about daily financial security, especially in communities that once supported agriculture or other industries, leave many with few alternatives.

For still others, the promise of the recent energy boom is often interpreted and given meaning through the lens of nationalism or through a belief that more oil means that hard choices about our consumption habits and the costs connected to them do not have to be examined closely. Several years ago, when I first taught my history of oil class at Rutgers, I asked the students whether oil was worth going to war over. One student, a veteran of the Iraq war, raised his hand and said yes. Conceding that oil was central to why the Iraq war was waged in the first place, he argued that war for oil is justified if it allows Americans to maintain their lifestyle. His candor was refreshing. He turned out to be one of my best students. He also changed his mind.

The argument that oil, and cheap oil in particular, is critical to the American way of life is another important element in why there is so much support for the current energy boom. Many Americans distrust foreign oil producers, particularly Arabs, who, they argue, support terrorism. These Americans instead support energy independence, which, as I argued in chapter 1, is a stubborn myth that has endured since the 1970s.

As I mentioned earlier, there has also been unexpected support from environmental groups that have historically advocated against long-term dependence on carbon energy. The Sierra Club, the largest and oldest organization advocating for environmental protection and sustainable practices in the United States, accepted over $25 million in donations from Aubrey McClendon, the CEO of one of the country's largest domestic energy companies, Chesapeake Energy, between 2007 and 2010. Bryan Walsh at *Time* magazine reported that the donations were directed toward the Sierra Club's Beyond Coal campaign. Like other environmental groups as well as domestic gas and oil advocates, the Sierra Club has argued that natural gas is a bridge fuel to an imagined future in which there are alternative sources to carbon. Until recently many believed that natural gas and even oil are cleaner sources of fuel than coal, which has long been considered a particularly terrible polluter and especially responsible for climate change. It turns out, however, that the belief in natural gas's cleaner footprint was mistaken. Methane leakage from gas wells is a much more significant contributor to the collection of dangerous greenhouse gases in the atmosphere than once thought. The Sierra Club turned away from the energy industry and its windfall support in 2010, when its new executive director, Michael Brune, announced its dramatic reversal: "It's time to stop thinking of natural gas as a 'kinder, gentler' energy source. What's more, we do not have an effective regulatory system in this country to address the risks that gas drilling poses on our health and communities."[8]

Big Energy, especially the new boom in natural gas production and its ability to generate big profits, has lured ostensibly liberal and environmental investors as well. George Soros, at least nominally committed to progressive and environmental causes, as evidenced by his support for groups like the Natural Resources Defense Council, invested in Consol Energy, a major coal and natural gas company, in 2014, though he soon sold off his investment. Even for high-profile figures who provide both rhetorical and material support for environmental protection, the payoff that comes from sacrificing principle for energy-related profit is often difficult to resist.

While the energy boom enjoys significant support, it has also come up against enthusiastic opposition. Across the country criticism has been especially sharp of the new wave of fracking and horizontal drilling. Some of this dissent has been highly visible. In 2010 Josh Fox, who lives near intense drilling activity along the Delaware River in Pennsylvania, confronted the energy rush in his polemical *Gasland* on HBO. In 2013 Fox released the sequel, *Gasland 2*. The films were devastating critiques of the domestic energy industry, dramatically bringing to light environmental and corporate abuse. Bill McKibben and large national anti-carbon-energy networks like 350.org, which are mostly concerned with averting catastrophic climate change, have kept criticism of the gas and coal industries alive in popular media outlets like *Rolling Stone* magazine. McKibben travels across the country, rallying university students and promoting an end to dependence on oil, coal, and gas.

The most meaningful opposition, however, has been local and less visible and has become the front line in the battle over community rights and democratic power. Local and community-based struggles have emerged in virtually every community where fracking has taken hold. This opposition has taken various forms: neighbors in parts of Pennsylvania, often at odds with one another on other political issues, organized after damage to their water wells or property, and community grassroots coordination has taken shape in Colorado, Texas, and elsewhere. Concerns about the energy industry's practices are based on several issues. Most important are anxieties about damage to property, public health, and the environment. Those most directly affected, the landowners and residents who live near fracking sites and well activity, were the earliest critics of industry practice and the dangers it constitutes. Largely ignored initially, partly because they were disparate, disorganized, and up against much more powerful political and financial forces, small networks of critics have gained momentum in recent years. Their concerns about threats to water, health, and other local environmental matters have been bolstered and legitimized by a growing scientific consensus that fracking does, in fact, come at significant expense to communities.

The energy industry has hardly backed down from the fight. Advocates have sought to counter criticism by mobilizing their massive resources to control the message. Attempts to dismantle Fox's *Gasland* critique, for instance, came from pro-energy organizations like Energy in Depth, which are funded by the industry. Other efforts include greenwashing, attempts to promote the energy industry as environmentally aware and committed to sustainable practices. Energy companies from Ohio to Colorado have quietly funded editorials in local and state newspapers, periodically passing off corporate flacks as simply concerned citizens. The *Denver Post*, Colorado's largest newspaper, created a section called "Energy and the Environment" in February 2014 that was sponsored by Coloradans for Responsible Energy Development, a group that supports natural gas development. The paper is transparent about the sources of support and that the content is driven by sponsors, not by journalists on staff.

More disturbingly, energy companies and advocacy groups have lavished cash on schools to bring in pro-fracking speakers and to promote programming that is sympathetic to the industry to unwitting young audiences. Talisman Energy, a Canadian company, even printed a twenty-four-page coloring book targeting children featuring Talisman Terry the Fracosaurus.[9] In late 2013 Radio Disney, funded by the pro-industry Ohio Oil and Energy Education Program, went on a tour of schools, community centers, and community events across Ohio promoting the oil and gas industry. A public petition compelled Disney Radio to cease the campaign. Good advertising, however misleading, boils down to being effective at public relations. But like tobacco companies that used to peddle their product to children, one should wonder about the ethics of branding carbon energy consumption to young people. While it is not illegal to do so, to sponsor school clubs or pay shills to write favorably on big energy's behalf has a dastardly and unethical quality. After all, it is one thing to offer arguments about the safety of and need for ongoing intense gas and oil extraction transparently and in whatever passes as a public marketplace of ideas. It is another thing altogether to use the industry's collective purchasing power secretly to buy favor and opportunity. If the industry believes its own claims that it is doing important and safe work, why resort to mystification and what amounts to public relations trickery?

Beyond manipulating advertising and the time-tested practice of selling snake oil, Big Energy engages in much more pernicious threats to community rights and has proven able to mobilize public authorities, sworn to serve and protect citizens, to do its bidding. In 2010 Pennsylvania State Police, the state's Office of Homeland Security, and Cabot Energy, one of the state's largest energy interests, coordinated to designate opposition to the energy company's operations as a potential terrorist threat. In his book *Under the Surface* Tom Wilber writes that opponents of Cabot Energy "had been listed on a bulletin [sent] to local law enforcement agencies and energy companies by the Pennsylvania Office of Homeland Security that detailed possible terrorist activity targeting key commercial facilities and energy infrastructure." State officials warned police that activists were tantamount to "anti-drilling environmentalist militants" who potentially sought to confront "gas drilling employees."[10]

Virginia Cody, an antifracking activist, was passed briefings and notes on her and her collaborators' activities from the state's security services. Wilber notes that Cody was stunned to learn that she and her supporters had been surveilled by the state and that the same Pennsylvania "intelligence network tracking al-Qaeda" was tracking them. Cody went public with her discovery and was reproached by Jim Powers, Pennsylvania's homeland security director, who, clearly not understanding that she would be no such partner, wrote, "Please assist us in keeping the information provided in the PIB [Pennsylvania Intelligence Bulletin] to those having a valid need-to-know; it should only be disseminated via closed communications systems. Thanks for your support. We want to continue providing this support to the Marcellus Shale Formation natural gas stakeholders while not feeding those groups fomenting dissent against those same companies."[11]

Pennsylvania's antifracking activists were marked for greater scrutiny by a private firm, the Institute for Terrorism Research and Response, which had a lucrative contract with the state to gather intelligence on potential terrorist threats. Wilber observed caustically that ITRR found terrorism "wherever it looked": "The list of events with the potential for terrorist activity, according to bulletins issued by ITRR, included a forestry industry conference, a screening of *Gasland*, several municipal zoning hearings regarding Marcellus Shale gas field development, [and] a political demonstration in favor of a tax on gas extraction."[12] Ultimately the Pennsylvania State Senate carried out an investigation and hearings into the linking of activism to terrorism. The fallout generated significant opprobrium, forcing Governor Ed Rendell to apologize publicly, cancel the ITRR contract, and disavow any knowledge of the firm's activities.

The investigation into state surveillance of antifracking and anti–Big Energy activists eventually resulted in some officials acknowledging citizens' rights to protest. To be sure, the democratic process worked in the end, with elected officials scrutinizing Pennsylvania intelligence services and successfully ending coordination between police and the corporate interests that were aligned against the community. But consider the scale of intervention and the accidental turn of events that created the conditions for such a reversal. Police and intelligence authorities defaulted not to their stated mission of protecting the community but to aligning with the power and influence of Big Energy.

Big Energy is hardly alone either historically or today in its ability to capture and harness the power of government. With the rise of modern industry and modern capitalism, political power has been easily influenced by what we euphemistically call "special interests." Sprawling banks along with the rest of Wall Street, Monsanto, large energy companies, and other megacorporations have successfully risen to prominence in part through their ability to finance legislation, sometimes even writing it themselves.

Powerful energy companies are perhaps exceptional in their ability to align not just with political elites but also with the police and the military. It is important to note that this alignment is a global phenomenon. The episode in Pennsylvania is not unusual in the United States or elsewhere. Energy companies everywhere have a history of close ties with the handful of agencies—police, intelligence, and military—that control the legitimate means of violence in the modern world. Some of the most oppressive and effective police states in the world are energy-rich regimes such as Iran, Nigeria, Saudi Arabia, and Russia. Energy companies offer what appear to be plausible justifications for these kinds of coordination: their infrastructure is sprawling and vulnerable, their operations are susceptible to interdiction or sabotage, and, given the intensity of some activism, special protection is necessary. It is not unreasonable for energy companies to demand and provide protection for facilities and infrastructure, of course. But what happened in Pennsylvania is closer to what happens in police states in the Global South, in which the interests of corporate energy and the state appear to be the same and state authorities have proven willing to define criminality and what counts as threatening in ways that favor industry rather than public order, the environment, or citizens.

How these arrangements come about is impossible to know. Perhaps they are settled quietly in backroom deals, and perhaps deal-making officials believe that the industry deserves special consideration. After all, supporting an industry that claims it is creating jobs and protecting the U.S. economy is deeply rooted in how we think about oil. The consequence, however, is that the energy industry ends up becoming a surrogate for the government.

Supporting Big Energy, then, is not just about making oil, gas, and usable products available for consumption, transportation, and industry. It is also about creating the conditions for certain kinds of state behavior in which the police are prepared to work in ways counter to their official responsibilities. It might be tempting to dismiss what happened in Pennsylvania as an anomaly in the democratic West. But in the past few years, clashes between antifracking activists and police have spiked, including in Canada, where Elsipogtog and Mi'kmaq First Nations tribes have been resisting government efforts to drill for oil on their lands, as well as in England and Romania. And these are only the activities we know about. Intelligence gathering and surveillance are, by definition, cloaked in secrecy and brought to light only by whistleblowing, which comes with significant risks.

Aside from the surveillance of activists, political authorities have helped shield energy interests from too much public scrutiny. In April 2013, after over five thousand barrels of oil spilled in Mayfair, Arkansas, a relatively small spill considering other catastrophes, the Federal Aviation Administration declared a no-fly zone over the area and gave exclusive flyover rights to Exxon, owner of the faulty pipeline. Exxon declared that it needed the rights in order to provide air-directed coordination to cleanup crews. While Exxon certainly should have been granted access to mobilize its resources, it was an odd decision on the part of the FAA to shut down the entire area to media concerned about documenting the scale of the spill and observing Exxon's remediation efforts.

The FAA, at the request of the U.S. Coast Guard, imposed a similar no-fly zone over the Gulf of Mexico after the BP Deepwater Horizon spill in 2010. U.S. authorities claimed that the decision was theirs and intended to minimize traffic and disruption of cleanup efforts. This may have been generally true, but the chain of authority was not always clear. A CBS TV crew, threatened with arrest when it tried to get access to oil-soaked beaches, was told by authorities that these were "BP's rules, not ours." In reporting on the media blockade, *Newsweek* magazine noted, "The problem, as many members of the press see it, is that even when access is granted, it's done so under the strict oversight of BP and Coast Guard personnel."[13] Whatever the intent, the FAA's and Coast Guard's decision enabled BP to determine the terms of coverage, a clear conflict of interest. It is unlikely in this case that the alignment of political and corporate power was malicious, although it was curious. Still, the result was deference to Big Energy, which, as details of the spill and the scale of damage would later reveal, deserved no special consideration.

Even when public officials challenge the activities and practices of Big Energy, as when Pennsylvania state senators pushed back against state intelligence authorities for surveilling antifracking activists, most continue to support the industry itself. At both the local and national level, almost every public official enthusiastically supports the recent energy boom. In his January 2014 State of the Union address President Obama

suggested that "if extracted safely," gas could be "the 'bridge
fuel' that can power our economy with less of the carbon pollution that causes climate change."[14] The White House's interest in bridge fuels stems partly from anxieties about how best
to address climate change without harming the energy industry. It also reflects the belief that the United States should and
can become energy independent, that energy is central to the
U.S. economy, and that there are no reasonable alternatives in
the short term. With regard to the economy, the administration has argued that given the challenges that have troubled
the country since the 2008 financial collapse, particularly high
unemployment, the energy industry is important because it is
an engine for jobs. Obama has argued that "one of the biggest factors in bringing more jobs back is our commitment to
American energy . . . and today, America is closer to energy
independence than we've been in decades."[15]

Across the country, state officials have similarly embraced
the dual claims that the energy industry is important for its role
in saving the climate and for creating jobs and invigorating
state economies. Governor Hickenlooper of Colorado is a particularly devoted gas and oil advocate. A former energy industry geologist, Hickenlooper has close ties to and outspoken
admiration and unyielding support for the industry. He frames
his support in terms similar to the president's. In his 2013 Colorado State of the State address he argued, "Colorado's economic welfare depends on how effective we are in developing
all of our resources. Our physical welfare requires we protect
public health and safety as we develop these resources. We can
reduce carbon emissions [and] create good-paying jobs." Echo-
ing the call for energy independence, he claimed, "Because of
innovations in drilling technology, cheaper, abundant natural
gas is helping to make America energy secure."[16]

Critics of these arguments have significant reasons to be
skeptical. The recent energy boom has certainly created jobs,

but it is difficult to know how sustainable this growth is in the long term. After all, part of the initial appeal of turning to oil and gas exploration was that it was not particularly labor-intensive compared to mining coal. While there has been significant job growth in the energy industry in the past few years, there are no assurances that this will last. Because fracking in shale and other deep geological formations is very expensive, requiring high oil and gas prices to be profitable and sustainable, shocks to global oil markets or rapid reductions in prices will likely limit growth, production, and the number of jobs available. Such a price reduction took place in late 2014, when prices collapsed over 40 percent between June and December. Perhaps most important, there is evidence that the green energy industry, which explores non-carbon-based alternative sources of fuel and power, is probably an even better engine of longer-term job creation. In 2012 the U.S. Energy Department reported that the green energy industry had created thousands of jobs across the country and could easily generate tens of thousands more if the industry was better supported.[17]

Aside from whether or not drilling for oil and gas is a sustainable model for economic development and growth, there is also significant evidence that natural gas, while a less dangerous emitter of greenhouse gases, is not better than oil for the climate. In October 2014 a group of researchers published a piece in the leading scientific journal *Nature* in which they argued, "Our results show that although market penetration of globally abundant gas may substantially change the future energy system, it is not necessarily an effective substitute for climate change mitigation policy."[18]

The economic benefits of the energy boom and the fracking rush have been greatest for the industry itself. For the overall economy it is difficult to determine how meaningful the impact of domestic energy production will be; for example, the falling oil prices of 2014 had very little to do with domestic production. Yet national and state leaders have not been persuaded that environmental risks associated with the boom are significant enough to warrant closer scrutiny. In 2013 the White House called for safe practices on the part of the domestic natural gas industry but refused to force it to disclose the chemical makeup of the water used in hydraulic fracturing. Governor Hickenlooper has been hostile to greater oversight of the energy industry, remarking in 2013 that "a patchwork of rules and regulations" would harm both the industry and the economy, no matter that more oversight would better serve the environment and public health. Both Obama and Hickenlooper are trusting that the economic upside of the energy boom, as yet unproven, trumps the environmental and health risks that the industry's practices create.

Activists in Colorado have pushed back against both the energy industry and its political backers. Not persuaded by the argument that the ostensible economic upside is worth the environmental risks, many in Colorado have pursued community-based measures to challenge the boom. Because state authorities, including the governor, have thrown their full support behind the energy companies, the struggle between citizens, activists, and political elites has put the very nature of democratic politics at the center of Colorado's struggle.

The scale of the challenge is significant, as Colorado's political structure and the legal system in which property rights are determined favor the energy industry. As is the case in many states, Colorado's landownership laws distinguish between surface rights and mineral rights, a system called "split estate." Homeowners and others who purchase land rarely own the mineral rights underground, particularly the rights to oil and gas. The U.S. Department of the Interior explains that this puts surface owners at a significant legal disadvantage because "mineral rights are considered the dominant estate, meaning they take precedence over other rights associated with the property, including those associated with owning the surface."[19] Since the beginning of the energy boom, energy companies have bought mineral rights across the state, especially along the Front Range, giving them the right to drill and explore for oil and gas over objections from landowners. Tens of thousands of wells have been drilled in the past few years, layering the eastern edge of the Rocky Mountains with a dense network of well sites and underground infrastructure. Residents across the state have complained that drilling companies appear on their property, build significant drilling and wastewater facilities, and operate noisily and dangerously twenty-four hours a day, often in close proximity to homes. There are vague expectations that drillers will not harm property, will act ethically, and will abide by minimal restrictions and setbacks (not drilling closer than five hundred feet from residential structures or schools). But the most significant source of alarm is that drilling often affects water and other resources.[20]

Governor Hickenlooper and other Colorado officials have consistently taken the side of the energy industry, refusing or moving only slowly to address citizens' anxieties. The alignment between high-level state authorities and the energy industry is significant. State officials have tremendous influence in shaping energy and environmental policy because Colorado's state constitution allows them rather than local authorities to make decisions about environmental management, resource extraction, and whether or not the energy industry is accountable to state or local officials. In Boulder County, to note one particularly important example, is a statutory community; county commissioners are elected by local residents but are nevertheless required to defer to the interest of the state and not those who voted for them. In this scenario, voters who oppose local industrial and drilling practices have faced challenges in putting their preferred policies into place, especially around agriculture and energy.

Figure 6. A fracking well on a Colorado farm. Because of laws governing split estates, landowners cannot prevent energy companies from using their property to explore for oil and gas underground. Photograph by Sandy Russell Jones.

In late 2013 concern about the environmental and public health risks posed by the energy boom mobilized Boulder County residents to challenge both the industry and the political structure that favored it. Because property owners were unable to prevent drilling and fracking on individual bases, at least not well enough to dent the industry's progress, five townships—Boulder, Longmont, Fort Collins, Lafayette, and Broomfield—voted to ban fracking pending studies about its safety and risk. Except in Broomfield, where the community was more divided, the measures received overwhelming support, with approval for the bans ranging between 55 and 75 percent. One of the leading antifracking and democratic rights activists in Boulder County, Cliff Willmeng, who has been a watchdog of the industry and an elegant spokesperson on behalf of the community, commented that the voters were "saying that they don't buy the idea that corporate interests are superior to public health, property values, quality of life and democratic self-determination."[21]

In December the Colorado Oil and Gas Association, a pro-energy trade association, filed lawsuits against the fracking moratoriums in Fort Collins and Lafayette, claiming that the communities were ordered by Colorado's constitution to defer to state authorities, which trumps any attempt at local regulation. In July 2014 COGA filed a similar lawsuit against the town of Longmont, where the fracking ban had support of 60 percent of voters. This time COGA was joined by the state. It was the second time Colorado officials had signed on in support of a trade association lawsuit against communities seeking to regulate fracking. Matt Lepore, head of the Colorado Oil and Gas Conservation Commission, the state office responsible for overseeing the energy industry, remarked to a local journalist, "The COGCC does believe Longmont's ban on hydraulic fracturing is contrary to state law, and we believe clarity from the courts on this matter is important for all parties."[22]

Anticipating that the lawsuits would face trouble in Colorado's courts, which must abide by the state constitution, several networks of activists launched a campaign to wrestle local rights away from the state. Although Colorado's constitution places Boulder and other energy-rich communities along the Front Range under central state authority, it does provide a mechanism for local citizens to vote and pass a home-rule amendment during the two-year election cycle. Home rule would allow citizens and their representatives to make decisions that reflect their will rather than the will of central authorities. In order to place two home-rule initiatives on the November 2014 ballot, which would likely have passed overwhelmingly, activists and Boulder's national congressional representative Jared Polis launched a grassroots campaign that aimed to gather over eighty-six thousand needed signatures. By early August the signature campaign appeared well on its way to success.

Community efforts to counter the energy industry and the state suffered a series of significant setbacks from midsummer into the fall. In late July a Colorado District Court judge ruled in favor of the COGA lawsuit against Longmont, remarking in her decision that "Longmont does not have the authority, in a matter of mixed state and local concern, to negate the authority of the Commission [COGCC]" and that the community "does not have the authority to prohibit what the state authorizes and permits."[23] Similar decisions to strike down the fracking bans were handed down for Fort Collins and Lafayette. In addition to the industry and state's coordinated efforts to scuttle the bans in the courts, Governor Hickenlooper, this time with an unexpected partner, was also able to sabotage the home-rule ballot initiative.

On August 4 Hickenlooper and Congressman Polis, one of the original backers of the grassroots ballot initiative, struck a backroom deal to pull the home rule amendments and halt the signature drive. The deal generated significant heartbreak in the activist community, which had gathered over a quarter of a million signatures and saw Polis's decision as a betrayal. And it was. But it was also a carefully considered and difficult decision. In exchange for Polis's cooperation, state authorities passed some minor concessions, including pulling two pro-industry ballot initiatives that would have created even more space for energy companies to drill where and as they liked. The *Boulder Weekly* wrote that in exchange for the deal Polis got the state to agree to appoint "a 21-member taskforce made up of oil and gas industry insiders, mainstays of the Democratic Party loyal to the governor and citizens or representatives of environmental groups who support more regulation of the oil and gas industry and fracking. . . . The governor also agreed to drop the state's lawsuit against the City of Longmont for having established oil and gas regulations considered stricter than the state's and promised to enforce a 1,000-foot setback as the norm rather than the exception."[24] As of late 2014, the task force was still establishing a framework for negotiating community versus industry concerns.

In an October piece reflecting on the deal, the *Boulder Weekly* pointed out that while Polis had sold out local activists, his calculation for doing so reflected several realities, including the possibility that the ballot initiatives would have been defeated, dealing a potentially permanent blow to the movement that hopes to check energy industry abuse. The forces in support of the industry are powerful and wealthy and could have vastly overpowered the resources of the antifracking community. Just as important, the landscape in support of fracking is ideologically complicated, including many national and local members of the Democratic Party, the party to which Hickenlooper belongs, as well as those national environmental groups such as the Sierra Club that have in the past supported fracking for natural gas as an alternative to coal.

The political dealing and courtroom setbacks have come as a disappointment to those concerned about the privilege of the energy industry, but these are likely just temporary obstacles. Many in Boulder County and elsewhere are committed to continuing the struggle. It is important that they do so and especially that they continue to link environmental rights with democratic rights.

WHAT IS MOST IMPORTANT about developments in Pennsylvania and Colorado over the past decade is the degree to which political authorities have aligned with the interests of the energy industry rather than local communities. It is tempting to accept that Governor Hickenlooper and others really do believe that fracking is a more sustainable source of fuel and power, better for the environment, and an engine for local and national economies. It is foolish to do so, however, as there is little evidence that any of these objectives will be met by fracking. What is needed is an end to capitulation to either the industry or political leaders and a great deal more skepticism and demands for community protection and rights. Left to their own, the energy industry and its partners have demonstrated historically, globally, and currently that they will use extralegal measures as well as long-standing favorable political conventions and property laws to aggrandize themselves and to hollow out the rights of locals in the pursuit of profit. In addition to the damage done to the environment, then, of equal concern is that carbon-based energy continues to shape political systems to suit industry desires. Oil and natural gas are dangerous forces against democratic politics.

In spite of the remaining challenges in Pennsylvania and Colorado, the struggle to combat the industry's environmental and political excesses has had some success. In December 2014 New York's governor Andrew Cuomo issued an order banning fracking in the state, citing concerns that the potential economic benefits did not outweigh the environmental and health risks. New York's health commissioner Howard Zucker remarked that "the overall weight of the evidence . . . demonstrates that there are significant uncertainties about the kinds of adverse health outcomes that may be associated with [fracking], the likelihood of the occurrence of adverse health outcomes, and the effectiveness of some of the mitigation measures in reducing or preventing environmental impacts which could adversely affect public health."[25]

New York's decision to outlaw fracking will almost certainly lead to years of industry lawsuits. Even so, it was an inspiring move. It was also the outcome of years of activist mobilization across the state, the success of a resilient and persistent community rights network that was able to successfully challenge Big Energy.[26] New York's example shows that while antidemocratic Big Energy is powerful, it does not always win.

Notes

Chapter 1   Choosing Energy

1. Jim Efstathiou, "Radiation in Pennsylvania Creek Seen as Legacy of Fracking," Bloomberg.com, October 2013.
2. Qtd. in ibid.
3. Susan Phillips, "Fracking's Other Danger: Radiation," stateimpact.npr.org, January 25, 2013.
4. Qtd. in ibid.
5. Permafix, *Technologically Enhanced Naturally Occurring Radioactive Materials (TENORM) Study Report*, prepared for Pennsylvania Department of Environmental Protection, January 2015, http://www.elibrary.dep.state.pa.us/dsweb/Get/Document-105822/PA-DEP-TENORM-Study_Report_Rev._Q_01-15-2015.pdf.
6. Adam Wernick, "As Fracking Booms, Waste Spills Rise—and So Do Arsenic Levels in Groundwater," www.pri.org, November 18, 2014, http://www.pri.org/stories/2014-11-18/fracking-booms-waste-spills-rise-and-so-do-arsenic-levels-groundwater.
7. Katie Valentine, "Nearly 3 Million Gallons of Drilling Waste Spill from North Dakota Pipeline," thinkprogress.org, January 22, 2015, http://thinkprogress.org/climate/2015/01/22/3614226/north-dakota-brine-spill/.
8. Alex Nussbaum, "Radioactive Waste Booms with Fracking as New Rules Mulled," Bloomberg.com, April 15, 2014.
9. Keith Matheny, "Michigan Landfill Taking Other States' Radioactive Fracking Waste," *Detroit Free Press*, freep.com, August 19, 204. See also Katie Valentine, "Michigan Is Taking the Radioactive Fracking Waste That Other States Rejected," thinkprogress.org, August 20, 2014.
10. Michael Barbaro, "Chris Christie Stumps for Energy, If Not for 2016, in Canada," *New York Times*, December 4, 2014.
11. Ibid.

12. Matthew Brown, "Montana City to Hear If Water Safe to Drink after Oil Spill," Associated Press, January 22, 2015.

13. Richard Nixon, "Special Message to Congress on Energy Resources," June 4, 1971, http://www.presidency.ucsb.edu/ws/?pid=3038.

14. Paul Sabin, "Crisis and Continuity in U.S. Oil Politics, 1965–1980," *Journal of American History*, June 2012, 179.

15. Richard Nixon, "Special Message to Congress on Energy Policy," April 18, 1973, http://www.presidency.ucsb.edu/ws/?pid=3817.

16. Ibid.

17. Joe Stork, *Middle East Oil and the Energy Crisis* (New York: Monthly Review Press, 1975).

18. Ibid.; Timothy Mitchell, *Carbon Democracy: Political Power in the Age of Oil* (New York: Verso, 2011).

19. Richard Nixon, "Address to the Nation about National Energy Policy," November 25, 1973, http://www.presidency.ucsb.edu/ws/?pid=4051.

20. Ibid.

21. Toby Craig Jones, "America, Oil, and War in the Middle East," *Journal of American History*, June 2012, 208–18.

22. "John McCain, Rally 09/05/08," https://www.youtube.com/watch?v=6EzHNApBdC4.

23. Barbaro, "Chris Christie Stumps for Energy."

24. "Obama Announces Plans to Achieve Energy Independence," washingtonpost.com, January 26, 2009.

CHAPTER 2   DANGEROUS WATER

1. Office of the Governor, State of Colorado, Executive Order D 2013–06, September 13, 2013.

2. Paul Danish, "The Flood of the Century? Not Exactly," *Boulder Weekly*, October 19, 2013.

3. See "Larimer County: Thousands of Homes Destroyed by the Flooding," 9News, September 17, 2013, http://archive.9news.com/rss/story.aspx?storyid=355521.

4. Charlie Brennan and John Aguilar, "Eight Days, 1,000-Year Rain, 100-Year Flood," *Daily Camera*, September 21, 2013, http://www.dailycamera.com/news/boulder-flood/ci_24148258/boulder-county-colorado-flood-2013-survival-100-rain-100-year-flood.

5. Joel Dyer, "What's in the Water," *Boulder Weekly*, September 19, 2013.

6. COGCC, *2013 Flood Response*, November 2013.

7. Ibid.

8. David Neslin, "Natural Gas Drilling: Public Health and Environmental Impacts," testimony before the U.S. Senate Commit-

tee on Environment and Public Works and Subcommittee on Water and Wildlife, April 12, 2011.

9. "Colorado Weighs Need for New Oil Rules after Flood," CBS Denver, February 7, 2014, denver.cbslocal.com/2014/02/07/ Colorado-weighs-need-for-new-oil-rules-after-flood/, accessed November 3, 2014.

10. Quoted in Tom Kenworthy, "Full Extent of Oil and Gas Spills from Colorado Floods Remains Unknown," Climate Progress, October 7, 2013, http://thinkprogress.org/person/tkenworthy/page/6/.

11. Neslin testimony.

12. "Is Fracking Safe? The Top 10 Controversial Claims about Natural Gas Drilling," *Popular Mechanics*, http://www.popularmechanics.com/science/energy/g161/top-10-myths-about-natural-gas-drilling-6386593/.

13. Sandra Postel, "Hormone Disrupting Chemicals Linked to Fracking Found in Colorado River," *National Geographic*, December 20, 2013.

14. Ibid.

15. Center for Biological Diversity, "Documents Reveal Billions of Oil Industry Wastewater Illegally Injected into Central California Aquifers," October 6, 2014, http://www.biologicaldiversity.org/news/press_releases/2014/fracking-10-06-2014.html.

16. Neela Banerjee, "Oil Companies Fracking into Drinking Water Sources, New Research Shows," *Los Angeles Times*, August 12, 2014.

17. Robert W. Howarth, Renee Santoro, and Anthony Ingraffea, "Methane and the Greenhouse-Gas Footprint of Natural Gas from Shale Formations: A Letter," *Climatic Change*, May 2011.

18. Alan Neuhauser, "Respiratory, Skin Problems Soar Near Gas Wells, Study Says," usnews.com, September 10, 2014.

19. Robert Neuhauser, "Toxic Chemicals, Carcinogens Skyrocket Near Fracking Sites," usnews.com, October 30, 2014.

20. Lisa Mackenzie et al., "Birth Outcomes and Maternal Residential Proximity to Natural Gas Development in Rural Colorado," *Environmental Health Perspectives*, January 2014.

21. Quoted in Renee Lewis, "New Study Links Fracking to Birth Defects in Heavily Drilled Colorado," america.aljazeera.com, January 30, 2014.

22. Jessica Goad, "Colorado Governor John Hickenlooper Appears in Fracking Ad," thinkprogress.org, February 29, 2012, http://thinkprogress.org/climate/2012/02/29/434994/colorado-governor-john-hickenlooper-appears-in-fracking-ad/.

23. Colorado Oil and Gas Association website, http://www.coga.org/index.php/Hydraulic%20Fracturing_Policy#sthash.otSia-BEH.dpbs.

24. Robert N. Proctor, *Golden Holocaust: Origins of the Cigarette Catastrophe and the Case for Abolition* (Berkeley: University of California Press, 2012).
25. Environmental Protection Agency, Safe Water Drinking Act, http://water.epa.gov/lawsregs/rulesregs/sdwa/index.cfm.
26. Abrahm Lustgarten, "Injection Wells: The Poison beneath Us," ProPublica, June 21, 2012.
27. Environmental Protection Agency, Office of Water, *U.S. EPA's Program to Regulate the Placement of Waste Water and Other Fluids Underground*, EPA 816-F-04-040, June 2004, http://water.epa. gov/lawsregs/guidance/sdwa/upload/programregplaceunder.pdf.
28. U.S. Environmental Protection Agency, "Summary of the Resource Conservation and Recovery Act," November 12, 2014, http://www2.epa.gov/laws-regulations/summary-resource-conservation-and-recovery-act.
29. Joel Dyer and Jefferson Dodge, "America's Dirtiest Secret," *Boulder Weekly*, March 13, 2014.
30. Environmental Protection Agency, "Regulatory Determination for Oil and Gas and Geothermal Exploration, Development and Production Wastes," FRL-3403-9, 53 FR 25447, July 6, 1988.
31. Lustgarten, "Injection Wells."
32. U.S. Environmental Protection Agency, "Regulatory Determination for Oil and Gas and Geothermal Exploration, Development and Production Wastes."
33. Ibid., 4.
34. Dyer and Jefferson, "America's Dirtiest Secret."
35. Kirk Johnson, "E.P.A. Links Tainted Water in Wyoming to Hydraulic Fracturing for Natural Gas," *New York Times*, December 8, 2011.

CHAPTER 3 ARABIA ON THE FRONT RANGE

1. For a range of views on the environmental impact of the oil industry historically and globally, see Brian Black, *Petrolia: The Landscape of America's First Oil Boom* (Baltimore: Johns Hopkins University Press, 2010); Myrna Santiago, *The Ecology of Oil: Environment, Labor, and the Mexican Revolution, 1900–1938* (Cambridge, UK: Cambridge University Press, 2009); Ike Okonta et al., *Where Vultures Feast: Shell, Human Rights, and Oil* (New York: Verso, 2003).
2. Timothy Mitchell, *Carbon Democracy: Political Power in the Age of Oil* (New York: Verso, 2011).
3. Thomas G. Andrews, *Killing for Coal: America's Deadliest Labor War* (Cambridge, MA: Harvard University Press, 2010).
4. Terry Lynn Karl, *The Paradox of Plenty: Oil Booms and Petro-States* (Berkeley: University of California Press, 1997).
5. Martin Sandbu, "The Iraqi Who Saved Norway from Oil," *Financial Times Magazine*, August 29, 2009, http://www.ft.com/

intl/cms/s/0/99680a04-92a0-11de-b63b-00144feabdc0.html#axzz3TQGgNPOB.

6. Robert Vitalis, *America's Kingdom: Mythmaking on the Saudi Oil Frontier* (Stanford: Stanford University Press, 2006).

7. Toby Craig Jones, *Desert Kingdom: How Oil and Water Forged Modern Saudi Arabia* (Cambridge, MA: Harvard University Press, 2010).

8. Brian Walsh, "Exclusive: How the Sierra Club Took Millions from the Natural Gas Industry—and Why They Stopped," *Time*, February 2, 2012, http://science.time.com/2012/02/02/exclusive-how-the-sierra-club-took-millions-from-the-natural-gas-industry-and-why-they-stopped/.

9. Talisman Energy USA, *Talisman Terry's Energy Adventure*, 2010, http://old.post-gazette.com/pg/pdf/201106/201106talisman_coloringbook.pdf.

10. Tom Wilber, *Under the Surface: Fracking, Fortunes, and the Fate of the Marcellus Shale* (Ithaca, NY: Cornell University Press, 2012), xx.

11. Ibid., Kindle locations 3644–47.

12. Ibid., Kindle location 3652.

13. Matthew Philips, "Photographers Say BP Restricts Access to Oil Spill," *Newsweek*, May 25, 2010, http://www.newsweek.com/photographers-say-bp-restricts-access-oil-spill-72849.

14. "President Barack Obama's State of the Union Address," January 28, 2014, http://www.whitehouse.gov/the-press-office/2014/01/28/president-barack-obamas-state-union-address.

15. Ibid.

16. "John Hickenlooper in 2013 Governor's State of the State Speeches," On the Issues, http://www.ontheissues.org/Archive/2013_State_John_Hickenlooper.htm.

17. U.S. Department of Energy, "The Clean Energy Economy Is Creating Jobs," Energy.gov, May 31, 2012, http://energy.gov/articles/clean-energy-economy-creating-jobs.

18. Haewon McJeon et al., "Limited Impact on Decadal-Scale Climate Change from Increased Use of Natural Gas," *Nature* 514, no. 7523 (2014), http://www.nature.com/nature/journal/v514/n7523/full/nature13837.html.

19. U.S. Department of Interior, Bureau of Land Management, "Split Estate," http://www.blm.gov/wo/st/en/prog/energy/oil_and_gas/best_management_practices/split_estate.html.

20. For more on the split estate and the health impacts of drilling and fracking so close to homes, see the documentary film *Split Estate*, http://www.splitestate.com/.

21. "Three of Four Colorado Cities Pass Fracking Ban but Similar Measures Fail in Two of Three Ohio Cities," *World Oil*,

November 6, 2013, http://www.worldoil.com/Three-of-four-Colorado-cities-pass-fracking-ban-but-similar-measures-fail-in-two-of-three-Ohio-cities.html.

22. John Tomasic, "State Joins Suit against Longmont Fracking Ban," *Colorado Independent*, July 11, 2013, http://www.colora-doindependent.com/128472/state-joins-suit-against-longmont-fracking-ban.

23. Joel Rosenblatt and Jennifer Oldham, "Longmont's Fracking Ban Tossed as Colorado Vote Looms," *Bloomberg Business*, July 25, 2014, http://www.bloomberg.com/news/articles/2014-07-25/longmont-s-fracking-ban-tossed-as-colorado-vote-looms.

24. Joel Dyer, Matt Cortina, and Elizabeth Miller, "Who Killed the Vote on Fracking?," *Boulder Weekly*, October 2, 2014, http://www.boulderweekly.com/article-13435-who-killed-the-vote-on-fracking.html.

25. Ibid.

26. Stephen Mufson, "Here's the Grassroots Political Story behind the New York Fracking Ban," Wonkblog, *Washington Post*, December 18, 2014, http://www.washingtonpost.com/blogs/wonkblog/wp/2014/12/18/heres-the-grassroots-political-story-behind-the-new-york-fracking-ban/.

ABOUT THE AUTHOR

TOBY CRAIG JONES IS an associate professor of history and the director of the Global and Comparative History master's degree program at Rutgers University. He is the author of *Desert Kingdom: How Oil and Water Forged Modern Saudi Arabia*.